CASE STUDIES IN
IMMUNOLOGY

A CLINICAL COMPANION

SECOND EDITION

CASE STUDIES IN
IMMUNOLOGY
A CLINICAL COMPANION

SECOND EDITION

Fred S. Rosen
Harvard Medical School

■

Raif S. Geha
Harvard Medical School

CB

CURRENT
BIOLOGY
PUBLICATIONS

GARLAND PUBLISHING
ALERE FLAMMAM
Taylor & Francis Group

Text Editor:	Eleanor Lawrence
Editorial Assistant:	Richard Woof
Copyeditor:	Bruce Goatly
Production Editor:	Emma Hunt
Indexer:	Liza Weinkove
Illustration and Layout:	Blink Studio, London

Distributors:

World: Garland Publishing, 19 Union Square West, New York, NY 10003, US.
Inside Japan: Nankodo Co. Ltd., 42-6, Hongo 3-Chome, Bunkyo-ku,
Tokyo 113, Japan.

ISBN 0 8153 3363 3 (paperback)

A catalog record for this book is available from the British Library.

Library of Congress Cataloging-in-Publication Data
Rosen, Fred S.
     Case studies in immunology: a clincial companion/
     Fred S. Rosen, Raif S. Geha. — 2nd ed.
         p.   cm.
     Includes index.
     ISBN 0-8153-3363-3 (pbk.)
     1. Clinical immunology- -Case studies. I. Geha, Raif S.
     II. Title.
     [DNLM: 1. Immunologic Diseases — case studies. 2. Immunity —
     genetics — case studies. 3. Allergy and Immunology — case studies.]
     [RC582.R67 1999]
     616.07'9- - dc21
                           98-46006
                            CIP

This book was produced using QuarkXpress 3.32 and Adobe Illustrator 7.0.1.

Printed in United States of America.

Published by Current Biology Publications, part of Elsevier Science London
Middlesex House, 34–42 Cleveland Street, London W1P 6LB, UK
and Garland Publishing, a member of the Taylor & Francis Group,
19 Union Square West, New York, NY 10003, US.

Preface for the second edition

The study of immunology provides a rare opportunity in medicine to relate the findings of basic scientific investigations to clinical problems. The case histories in this book are chosen with two purposes: to illustrate in a clinical context essential points about the mechanisms of immunity; and to describe and explain some of the immunological problems most often seen in the clinic. For this second edition, we have added 11 completely new cases that illustrate both recently discovered genetic immunodeficiencies and some more familiar and common diseases with interesting immunology. Fundamental mechanisms of immunity are illustrated with cases of genetic defects in the immune system, immune complex diseases, immune mediated hypersensitivity reactions and autoimmune and alloimmune diseases. These cases describe real events from case histories, largely drawn from the records of the Boston Children's Hospital and the Brigham and Women's Hospital. Names, places and time have been altered to obscure the identity of the patients described; all other details are faithfully reproduced. The cases are intended to help medical students and pre-medical students to learn and understand the importance of basic immunological mechanisms, and particularly to serve as a revision aid; but we hope and believe they will be useful and interesting to any student of immunology.

Each case is presented in the same format. The case history is preceded by basic scientific facts that are needed to understand the case history. The case history is followed by a brief summary of the disease under study. Finally there are several questions and discussion points that highlight the lessons learned from the case. These are not intended to be a quiz but rather to shed further light on what has been learned from the case.

We are very grateful to Dr. Robertson Parkman of the Los Angeles Children's Hospital for the MHC class II deficiency case, to Dr. Henri de la Salle of the Centre Regional de Transfusion Sanguine in Strasbourg, France for the MHC class I deficiency case and to Prof. Michael Levin of St. Mary's Hospital, London for the interferon-γ receptor deficiency case. We are also greatly indebted to our colleagues Drs. David Dawson, Susan Berman and Lawrence Shulman of the Brigham and Women's Hospital, to Dr. Razzaque Ahmed of the Harvard School of Dental Medicine, to Dr. Ernesto Gonzalez of the Massachusetts General Hospital and to Dr. Peter Newburger of the Department of Pediatrics of the University of Massachusetts for supplying case materials. Our colleagues in the Immunology Division of the Children's Hospital have provided invaluable service by extracting summaries of long and complicated case histories; we are particularly indebted to Drs. Lynda Schneider, Leonard Bacharier, Francisco A. Bonilla, Hans Oettgen and Jonathan Spergel in constructing several case histories. In the course of developing these chapters, we have been indebted for expert and pedagogic advice to Mark Walport, Jan Vilcek, George Miller, Ten Feizi, Fenella Woznarowska, Michael J. Colston, Anthony Segal, Peter Parham, Emil Unanue, Leslie Berg, Christopher Goodnow, Hugh Auchincloss, Anthony De Franco and John J. Cohen.

Eleanor Lawrence has spent many hours honing the prose as well as the content of the cases and we are very grateful to her for this. Emma Hunt, Richard Woof and Matthew McClements have painstakingly organized the text and figures and without their vital work this book would not have come into being. Miranda Robertson determined the format of the presentation of these cases and made them into a valuable pedagogical tool with her peerless editorial skills.

A note to the reader

The cases presented in this book have been ordered so that the main topics addressed in each case follow as far as possible the order in which these topics are presented in the fourth edition of *Immunobiology* by Charles A. Janeway Jr., Paul Travers, Mark Walport, and J. Donald Capra. However, inevitably many of the early cases raise important issues that are not addressed until the later chapters of *Immunobiology*. To indicate which sections of *Immunobiology* contain material relevant to each case, we have listed at the head of each case the topics covered in it. The color code follows the code used for the five main sections of *Immunobiology*: yellow for the introductory and method chapters, blue for the section on recognition of antigen, red for the development of lymphocytes, green for the activation and effector functions of lymphocytes, and purple for the response to infection and clinical topics.

Photographs

The following photographs have been reproduced with the kind permission of the journal in which they were originally published.

Case 1

Fig. 1.2 bottom panel from the *Journal of Experimental Medicine* 1972, **135**:200–219. By copyright permission of the Rockefeller University Press.

Case 5

Fig. 5.2 bottom left panel from *Diagnostic Immunopathology* (2nd edition), eds. R.B. Colvin, A.K. Bahn, and R.T. McCluskey. New York, Raven Press 1995, 246–247. © 1995, Raven Press.
Fig. 5.4 from *Immunology* (3rd edition) by Roitt et al, 1993. Mosby–Year Book Europe Limited, London, UK.

Case 7

Fig. 7.4 from the *Journal of Experimental Medicine* 1990, **172**:981–984. By copyright permission of the Rockefeller University Press.

Case 9

Fig. 9.5 from *Clinical immunology and*

Immunopathology 1975, **4**:174–188. © 1975, Academic Press.

Case 14

Fig. 14.6 from *Immunity* 1998, **9**:81–91. © 1998, Cell Press.
Fig. 14.5 from *Blood* 1986, **68**:1329–1332. © 1986, W.B. Saunders and Co.

Case 15

Fig. 15.1 bottom panels from the *Journal of Experimental Medicine* 1987, **169**:893–907. By copyright permission of the Rockefeller University Press.

Case 29

Fig. 29.4 left panel from *International Review of Experimental Pathology* 1986, **28**:45–78. © 1986, Academic Press.

Case 30

Fig. 30.4 from the *Journal of Allergy and Clinical Immunology* 1983, **71**:47–56. © 1983, Mosby.

CONTENTS

CASE 1 | Congenital Asplenia

The role of the spleen in immunity.

The adaptive immune response occurs mainly in the secondary lymphoid tissue—the lymph nodes, the gut-associated lymphoid tissue, and the spleen (Fig. 1.1). Pathogens and their secreted antigens are trapped in these tissues, and presented to the naive lymphocytes that constantly pass through. Microorganisms that enter the body through the skin or the lungs drain to regional lymph nodes where they stimulate an immune response. Microorganisms and food antigens that enter the gastrointestinal tract are collected in the gut-associated lymphoid tissue. Microbes that enter the bloodstream stimulate an immune response in the spleen.

Topics bearing on this case:
Circulation of lymphocytes through secondary lymphoid tissues
Toxoid vaccines
Hemagglutination tests
T-cell help in antibody response

Fig. 1.1 The distribution of lymphoid tissues in the body. Lymphocytes arise from stem cells in bone marrow, and differentiate in the central lymphoid organs (yellow)—B cells in bone marrow and T cells in the thymus. They migrate from these tissues through the bloodstream to the peripheral lymphoid tissues (blue)—the lymph nodes, spleen, and gut-associated lymphoid tissues such as tonsils, Peyer's patches, and appendix. These are the sites of lymphocyte activation by antigen. Lymphatics drain extracellular fluid as lymph through the lymph nodes and into the thoracic duct, which returns the lymph to the bloodstream by emptying into the left subclavian vein. Lymphocytes that circulate in the bloodstream enter the peripheral lymphoid organs, and are eventually carried by lymph to the thoracic duct where they re-enter the bloodstream.

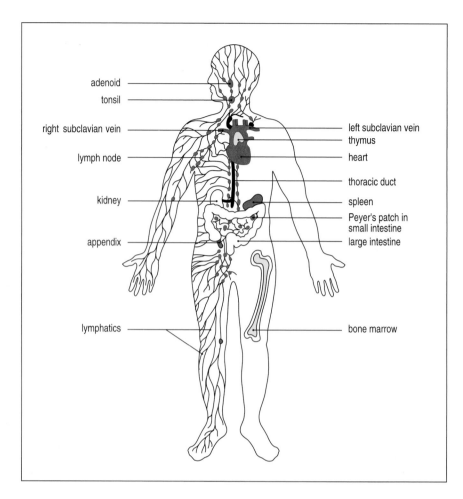

The spleen

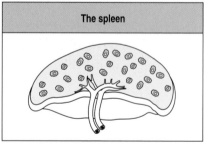

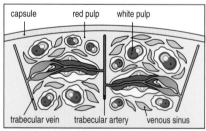

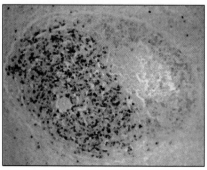

The spleen is organized to accomplish two functions (Fig. 1.2). In addition to being a secondary lymphoid organ, it acts as a filter of the blood to remove aged or abnormal red cells and other extraneous particles that may enter the bloodstream, including microorganisms. The lymphoid function of the spleen is carried out in the white pulp and the filtration function by the red pulp. Many microorganisms are recognized directly and engulfed by the phagocytes of the red pulp. Others are not removed efficiently until they are coated by antibodies generated in the white pulp. In experimental animals, an immune response (as measured by antibody formation) can be detected in the white pulp of the spleen approximately 4 days after the intravenous injection of a dose of microorganisms. The clearance of antibody- and complement-coated bacteria or viruses by the phagocytic cells of the red pulp of the spleen is very rapid. Rapid clearance from the blood is important as it prevents these bacteria from disseminating and causing infections of the meninges (meningitis), the kidney (pyelonephritis), the lung (pneumonia), or other distant anatomical sites.

Fig. 1.2 Schematic views and light micrograph of a section of spleen. The spleen consists of red pulp (pink areas), which is a site of red blood cell destruction, interspersed with lymphoid white pulp. The center panel shows an enlargement of a small section of the spleen showing the arrangement of discrete areas of white pulp around central arterioles. The white pulp is shown in transverse section. Although the organization of the spleen is similar to that of a lymph node, antigen enters the spleen from the blood rather than from the lymph. Photograph courtesy of J Howard.

Bacteria enter the bloodstream all the time, such as when we brush our teeth or when we have a local infection, for example of the skin or middle ear. Normally these bacteria are disposed of efficiently by the spleen. When, for one reason or another, the spleen is not present, serious, even fatal, infections occur.

The case of Susan Vanderveer: a fatality because of an absent spleen.

Mr and Mrs Vanderveer owned a farm in the Hudson Valley in lower New York State. They were both descended from Dutch settlers who came to the Hudson Valley in the mid 17th century. There were multiple consanguineous marriages among their ancestors, and Mr and Mrs Vanderveer were distantly related to each other. At the time of this case, they had five children—three girls and two boys. Their youngest daughter, Susan, was 10 months old when she developed a cold, which lasted for 2 weeks. On the 14th day of her upper respiratory infection, she became sleepy and felt very hot. Her mother found that her temperature was 41.7°C. When Susan developed convulsive movements of her extremities, she was rushed to the emergency room but she died on the way to the hospital. Post-mortem cultures of blood were obtained, and also from her throat and cerebrospinal fluid. All the cultures grew *Haemophilus influenzae*, type b. At autopsy Susan was found to have no spleen.

Susan Vanderveer, age 10 months, dead on arrival in Emergency.

At the time of Susan's death her 3-year-old sister, Betsy, also had a fever of 38.9°C. She complained of an earache and her eardrums were found to be red. She had no other complaints and no other abnormalities were detected on physical examination. Her white blood count was 28,500 cells μl^{-1} (very elevated). Cultures from her nose, throat, and blood grew out *Haemophilus influenzae*, type b. She was given ampicillin intravenously for 10 days in the hospital and was then sent home in good health. Her cultures were negative at the time of discharge from the hospital. She was seen by a pediatrician on three occasions during the following year for otitis media (inflammation of the middle ear), pneumonia, and mastoiditis (inflammation of the mastoid bone behind the ear).

Betsy Vanderveer, age 3 years, presents with severe H. influenzae infection.

David, Susan's 5-year-old brother, had been admitted to the hospital at 21 months of age with meningitis caused by *Streptococcus pneumoniae*. He had responded well to antibiotic therapy and had been discharged. Another occurrence of pneumococcal meningitis at 27 months of age had also been followed by an uneventful recovery after antibiotics. He had had pneumonia at age 3½ years. At the time of Susan's death he was well.

The two other children of the Vanderveers, a girl aged 8 years and a newborn male, were in good health.

All the Vanderveer children had received routine immunization at ages 3, 4 and 5 months with tetanus and diphtheria toxoids and killed *Bordetella pertussis* to protect against tetanus, diptheria, and whooping cough, which are three potentially fatal diseases caused by bacterial toxins (Fig. 1.3). Serum agglutination tests were used to test their antibody responses to these and other immunogens. Samples of serum from both Betsy and David caused hemagglutination (the clumping of red blood cells) when added to red blood cells (type O) coated with tetanus toxoid. Hemagglutinating antibodies to tetanus toxoid were seen at serum dilutions of 1:32 for both Betsy and David, and were found at a similar titer in their 8-year-old sister. All three children were given typhoid vaccine subcutaneously and 4 weeks later

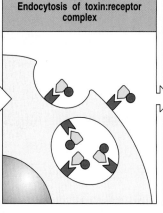

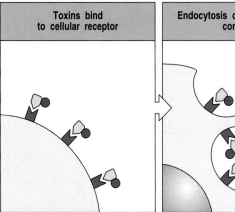

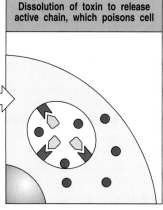

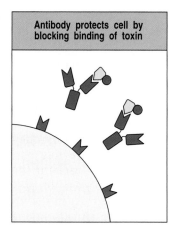

| Toxins bind to cellular receptor | Endocytosis of toxin:receptor complex | Dissolution of toxin to release active chain, which poisons cell | Antibody protects cell by blocking binding of toxin |

Fig. 1.3 Neutralization by antibodies protects cells from toxin action. Secreted bacterial toxins usually contain several distinct moieties. One piece of the toxin must bind a cellular receptor, which allows the molecule to be internalized. A second part of the toxin molecule then enters the cytoplasm and poisons the cell. In some cases, a single molecule of toxin can kill a cell. Antibodies that inhibit toxin binding can prevent, or neutralize, these effects. Protective antibodies can be generated by immunizing subcutaneously with toxoids. Toxoids are toxins rendered harmless by treating with denaturing agents, such as formalin, which destroy their toxicity but not their ability to generate neutralizing antibodies. In the case of the DPT vaccine, the killed *Bordetella pertussis* cells act as an adjuvant, which enhances the immune response to all components of the vaccine by delivering activating signals to antigen-presenting cells.

Agglutination tests to diphtheria, tetanus and pertussis toxins normal.

samples of their sera were tested for the ability to agglutinate killed *Salmonella typhosa*. The results indicated a normal immune response. David had an agglutination titer of 1:16, Betsy 1:32 and their normal 8-year-old sister 1:32. All three children were given 1 ml of a 25% suspension of sheep red cells intravenously. David had a titer of 1:4 for hemagglutinating antibodies against sheep red blood cells prior to the injection. He was tested again 2 and 4 weeks later and there was no increase in titer. Betsy had an initial titer of 1:32 and her titer did not rise either. The 8-year-old normal sister had a pre-immunization titer of 1:32. She was tested 2 and 4 weeks after the immunization, when she was found to have a hemagglutinating titer of 1:256 against sheep red blood cells.

All the children and their parents were injected intravenously with radioactive colloidal gold (Au^{198}), which is taken up by the reticuloendothelial cells of the liver and spleen within 15 minutes after the injection. A scintillation counter then scans the abdomen for radioactive gold. The pattern of scintillation reveals that Betsy and David have no spleens (Fig. 1.4).

Asplenia and splenectomy.

The genetic defect causing asplenia has not yet been identified. The Vanderveer family is unusual in that three of their first four children were born without spleens. After the events described in this case, the Vanderveers had three more children. One of the boys and the girl were also born without spleens; the other boy had a normal spleen. This family provides us with an uncomplicated circumstance in which to examine the role of the spleen. The major consequence of its absence is a susceptibility to bacteremia, usually caused by the encapsulated bacteria *Streptococcus pneumoniae* or *Haemophilus influenzae*. This susceptibility is caused by a failure of the immune response to these common extracellular bacteria when they enter the bloodstream.

Surgical removal of the spleen is quite common. The capsule of the spleen may rupture from trauma, for example in an automobile accident. In such cases, the spleen has to be surgically removed very quickly because of blood loss into the abdominal cavity. The spleen may also be removed surgically for therapeutic reasons in certain autoimmune diseases, or because of a malignancy in the spleen. After splenectomy, patients, particularly children, are susceptible to bloodstream infections by microorganisms to which they have no antibodies. Microorganisms to which the host has antibodies are removed quickly from the bloodstream by the liver, where the Kupffer cells complement the role of the red pulp of the spleen. Antibodies to the encapsulated bacteria that commonly cause trouble with bloodstream infections persist for a very long time in the bloodstream of exposed individuals, even in the absence of a spleen (for reasons that are not fully understood). Adults who already have antibodies to these microorganisms are therefore much less vulnerable to problems of bacteremia than children who have not yet developed antibodies to these germs.

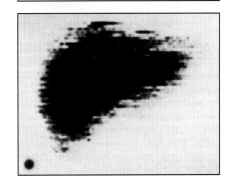

Discussion and questions.

1 *Nicholas Biddleboy, a 5-year-old boy, had his spleen removed following a sledding accident, during which both he and his sled struck a tree trunk. In the emergency room of a nearby hospital, it was determined that his spleen had ruptured. The surgeon, following removal of a spleen that had indeed ruptured, calls you for an immunology consultation. What do you advise?*

First you find out that Nicholas has had all his routine immunizations. He received DPT (diphtheria, pertussis, and tetanus antigens) and oral live poliovirus vaccine at ages 3, 4, and 5 months, and a booster of both before entering kindergarten. He was also given MMR (mumps, measles and rubella live vaccines) at 9 months of age. At the same time, he was given Hib vaccine (the conjugated capsular polysaccharide of *Haemophilus influenzae*, type b; Fig. 1.5). His growth and development have been normal. He suffered a middle ear infection (otitis media) at age 24 months. Other than that he has had no other illnesses, except for a common cold each winter. You feel comfortable that he is protected against infection with *Haemophilus influenzae* from the Hib vaccine. However, your concern about the possiblity of pneumococcal infection leads you to advise the surgeon to immunize Nicholas against pneumococcal capsular poysaccharides by giving him Pneumovax (a vaccine containing the major prevalent pneumococcal polysaccharides). You also advise prophylactic antibiotics, to be taken at a low dose daily but at higher doses when Nicholas has any dental work done, or any invasive surgical procedure.

2 *Why did David and Betsy have normal responses to the typhoid vaccine but not to the sheep red blood cells?*

The typhoid vaccine was given subcutaneously and a response was mounted in a regional lymph node. The sheep red blood cells were given intravenously, and, in the absence of a spleen, failed to enter any secondary lymphoid tissue where an immune response could occur.

Fig. 1.4 A scintillation scan of the abdomen after intravenous injection with radioactive colloidal gold (Au^{198}) reveals that Betsy and David Vanderveer have no spleens. The top panel shows an abdominal scan of Betsy's mother. The large mass on the left is the liver and the small mass on the right is the spleen. The reticuloendothelial cells of both liver and spleen take up the labeled gold within 15 minutes after the injection. No spleen is seen in either Betsy (middle panel) or David (lower panel).

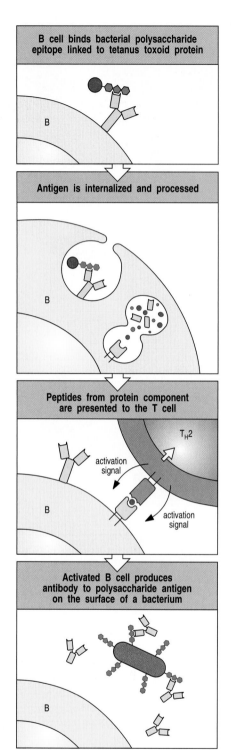

B cell binds bacterial polysaccharide epitope linked to tetanus toxoid protein

Antigen is internalized and processed

Peptides from protein component are presented to the T cell

T_H2

activation signal

activation signal

Activated B cell produces antibody to polysaccharide antigen on the surface of a bacterium

Fig. 1.5 *Haemophilus influenzae* b vaccine is a conjugate of bacterial polysaccharide with the tetanus toxoid protein, which enhances the immune response by allowing a polysaccharide-specific B cell to recruit T-cell help. The B cell recognizes and binds the polysaccharide, internalizes and degrades the toxoid protein to which it is attached, and displays peptides derived from it on surface MHC class II molecules. Helper T cells generated in response to earlier vaccination against the toxoid recognize the complex on the B-cell surface and activate the B cell to produce antibody against the polysaccharide. This antibody can then protect against *H. influenzae* b infection.

3 *The Vanderveer family is unique in the medical literature. The parents, who were distantly related, were normal and had normal spleens. Five of their eight children were born without spleens. Of these, only Betsy subsequently had children—four boys and one girl. They are all normal and have spleens. What is the inheritance pattern of congenital asplenia in this family? According to Mendelian laws how many of the eight Vanderveer children would be expected to have no spleen?*

The defect is inherited as an autosomal recessive. The parents are normal but each carries this recessive gene. Furthermore, they are consanguineous, a setting in which autosomal recessive disease is encountered more frequently than in outbred people. Chance would predict that one in four (that is, two) of their eight children would be affected. Each pregnancy provides a one in four chance of the fetus' inheriting the abnormal gene from both parents. As it turned out, this happened in five of Mrs Vanderveer's eight pregnancies. Since Betsy married a normal man, all her children are heterozygous for the defect, like their maternal grandparents, and have normal spleens.

CASE 2 | X-Linked Agammaglobulinemia

An absence of B lymphocytes.

Topics bearing on this case:

B-cell maturation

Effector functions of antibodies

Humoral versus cell-mediated immunity

Effector mechanisms of humoral immunity

Methods for measuring T-cell function

Actions of complement and complement receptors

One of the most important functions of the adaptive immune system is the production of antibodies. It is estimated that a human being can make over one million different specific antibodies. This remarkable feat is accomplished through a complex genetic program carried out by B lymphocytes and their precursors in the bone marrow (Fig. 2.1). Every day about 2.5 billion (2.5×10^9) early B-cell precursors (pro-B cells) take the first step in this genetic program and enter the body pool of pre-B cells. From this pool of rapidly dividing pre-B cells 30 billion daily mature into B cells, which leave the bone marrow as circulating B lymphocytes, while 55 billion fail to mature successfully and undergo programmed cell death. This process continues throughout life, although the numbers gradually decline with age.

Fig. 2.1 The development of B cells proceeds through several stages marked by the rearrangement of the immunoglobulin genes. The bone marrow stem cell that gives rise to the B-lymphocyte lineage has not yet begun to rearrange its immunoglobulin genes; they are in germline configuration. The first rearrangements of D gene segments to J_H gene segments occur in the early pro-B cells, generating late pro-B cells. In the late pro-B cells, a V_H gene segment becomes joined to the rearranged DJ_H, producing a pre-B cell that is expressing both low levels of surface and high levels of cytoplasmic μ heavy chain. Finally, the light-chain genes are rearranged and the cell, now an immature B cell, expresses both light chains (L chains) and μ heavy chains (H chains) as surface IgM molecules. Cells that fail to generate a functional surface immunoglobulin, or those with a rearranged receptor that binds a self-antigen, die by programmed cell death. The rest leave the bone marrow and enter the bloodstream.

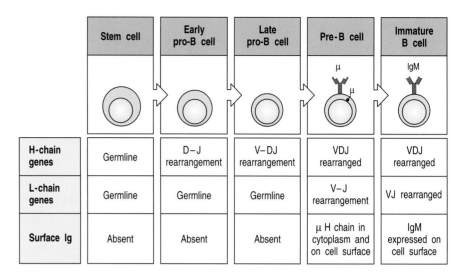

	Stem cell	Early pro-B cell	Late pro-B cell	Pre-B cell	Immature B cell
H-chain genes	Germline	D–J rearrangement	V–DJ rearrangement	VDJ rearranged	VDJ rearranged
L-chain genes	Germline	Germline	Germline	V–J rearrangement	VJ rearranged
Surface Ig	Absent	Absent	Absent	μ H chain in cytoplasm and on cell surface	IgM expressed on cell surface

Mature circulating B cells proliferate on encounter with antigen and differentiate into plasma cells, which secrete antibody. Antibodies, which are made by the plasma cell progeny of B cells, protect by binding to and neutralizing toxins and viruses, by preventing the adhesion of microbes to cell surfaces and, after binding to microbial surfaces, by fixing complement and thereby enhancing phagocytosis and lysis of pathogens (Fig. 2.2).

This case concerns a young man who has an inherited inability to make antibodies. His family history reveals that he has inherited this defect in antibody synthesis as an X-linked recessive abnormality. This poses an interesting puzzle because the genes encoding the structure of the immunoglobulin polypeptide chains are encoded on autosomal chromosomes and not on the X chromosome. Further inquiry reveals that he has no B cells, so that some gene on the X chromosome is critical for the normal maturation of B lymphocytes.

The case of Bill Grignard: a medical student with scarcely any antibodies.

Bill Grignard was well for the first 10 months of his life. In the next year he had pneumonia once, several episodes of otitis media (inflammation of the middle ear) and on one occasion developed erysipelas (streptococcal infection of the skin) on his right cheek. These infections were all treated successfully with antibiotics but it seemed to his mother, a nurse, that he was constantly on antibiotics.

His mother had two brothers who had died 30 years prior to Bill's birth from pneumonia in their second year of life, before antibiotics were available. She also has two sisters who are well; one has a healthy son and daughter and the other a healthy daughter.

Bill was a bright and active child who gained weight, grew, and developed normally but he continued to have repeated infections of the ears and sinuses and twice again had pneumonia. At 2 years and 3 months his local pediatrician tested his serum immunoglobulins. He found 80 mg dl^{-1} IgG (normal 600–1500 mg dl^{-1}), no IgA (normal 50–125 mg dl^{-1}), and only 10 mg dl^{-1} IgM (normal 75–150 mg dl^{-1}).

Two-year-old boy, two maternal uncles died in infancy from infection.

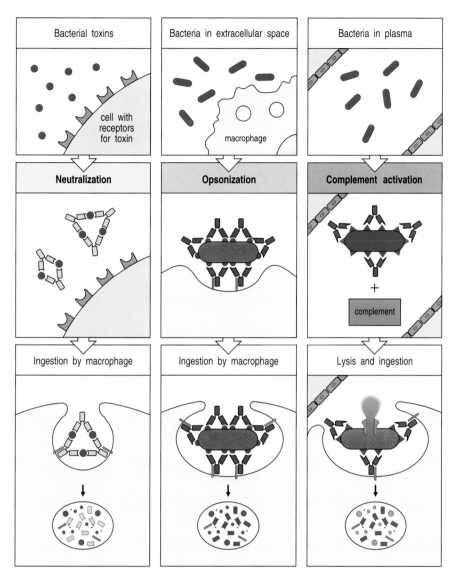

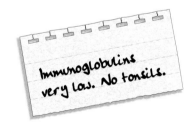

Fig. 2.2 Antibodies can participate in host defense in three main ways. The left column shows antibodies binding to and neutralizing a bacterial toxin, preventing it from interacting with host cells and causing pathology. Unbound toxin can react with receptors on the host cell, whereas the toxin:antibody complex cannot. Antibodies also neutralize complete virus particles and bacterial cells by binding to them and inactivating them. The antigen:antibody complex is eventually scavenged and degraded by macrophages. Antibodies coating an antigen render it recognizable as foreign by phagocytes (macrophages and polymorphonuclear leukocytes), which then ingest and destroy it; this is called opsonization. The central column shows the opsonization and phagocytosis of a bacterial cell. The right column shows the activation of the complement system by antibodies coating a bacterial cell. Bound antibodies form a receptor for the first protein of the complement system, which eventually forms a protein complex on the surface of the bacterium that favors its uptake and destruction by phagocytes and can, in some cases, directly kill the bacterium. Thus, antibodies target pathogens and their products for disposal by phagocytes.

Bill was started on monthly intramuscular injections of gamma globulin; his serum IgG level was maintained at 200 mg dl⁻¹. He started school at age 5 years and performed very well (he was reading at second grade level at age 5 years) despite prolonged absences because of recurrent pneumonia and other infections.

At 9 years of age he was referred to the Children's Hospital because of atelectasis (collapse of part of a lung) and a chronic cough. On physical examination he was found to be a well-developed, alert boy. He weighed 33.5 kg and was 146 cm in height (this height and weight is normal for his age). The doctor noted that he had no visible tonsils (he had never had a tonsillectomy). With a stethoscope the doctor also heard rales (moist crackles) at both lung bases.

Further family history revealed that Bill had one younger sibling, John, a 7-year-old brother, who also had contracted pneumonia on three occasions. John had a serum IgG level of 150 mg dl⁻¹.

Laboratory studies at the time of Bill's visit to the Children's Hospital gave a white blood cell count of 5100 μl⁻¹ (normal) of which 45% were neutrophils (normal), 43% were lymphocytes (normal), 10% were monocytes (elevated) and 2% were eosinophils (normal).

Fig. 2.3 The FACS™ allows individual cells to be identified by their cell-surface antigens and to be sorted. Cells to be analyzed by flow cytometry are first labeled with fluorescent dyes (top panel). Direct labeling uses dye-coupled antibodies specific for cell-surface antigens (as shown here), while indirect labeling uses a dye-coupled immunoglobulin to detect unlabeled cell-bound antibody. The cells are forced through a nozzle in a single-cell stream that passes through a laser beam (second panel). Photomultiplier tubes (PMTs) detect the scattering of light, which is a sign of cell size and granularity, and emissions from the different fluorescent dyes. This information is analyzed by computer (CPU). By examining many cells in this way, the number of cells with a specific set of characteristics can be counted and levels of expression of various molecules on these cells can be measured. The lower part of the figure shows how these data can be represented, using the expression of two surface immunoglobulins, IgM and IgD, on a sample of B cells from a mouse spleen. The two immunoglobulins have been labeled with different-colored dyes. When the expression of just one type of molecule is to be analyzed (IgM or IgD), the data are usually displayed as a histogram, as in the left-hand panels. Histograms display the distribution of cells expressing a single measured parameter (eg size, granularity, fluorescence color). When two or more parameters are measured for each cell (IgM and IgD), various types of two-color plot can be used to display the data, as shown in the right-hand panel. All four plots represent the same data. The horizontal axis represents intensity of IgM fluorescence and the vertical axis the intensity of IgD fluorescence. Two-color plots provide more information than histograms; they allow recognition, for example, of cells that are 'bright' for both colors, 'dull' for one and bright for the other, dull for both, negative for both, and so on. For example, the cluster of dots in the extreme lower left portions of the plots represents cells that do not express either immunoglobulin, and are mostly T cells. The standard dot plot (upper left) places a single dot for each cell whose fluorescence is measured. It is good for picking up cells that lie outside the main groups but tends to saturate in areas containing a large number of cells of the same type. A second means of presenting these data is the color dot plot (lower left), which uses color density to indicate high-density areas. A contour plot (upper right) draws 5% 'probability' contours, with 5% of the cells lying between each contour providing the best monochrome visualization of regions of high and low density. The lower right plot is a 5% probability contour map that also shows outlying cells as dots.

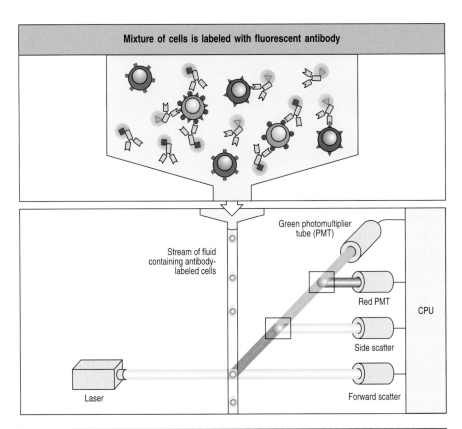

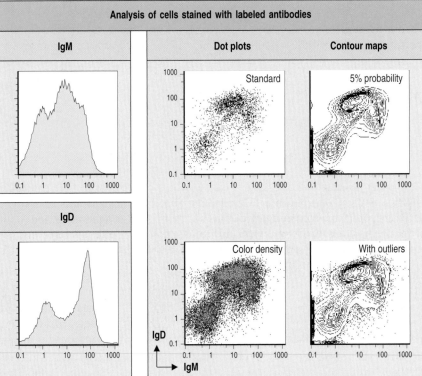

Flow cytometry (Fig. 2.3) showed that 85% of the lymphocytes bound an antibody to CD3, a T-cell marker (normal); 55% were helper T cells reacting with an anti-CD4 antibody; and 29% were cytotoxic T cells reacting with an anti-CD8 antibody (normal). However, none of Bill's peripheral blood lymphocytes bound an antibody to the B-cell marker CD19 (normal 12%) (Fig. 2.4).

Fig. 2.4 Clinical FACS analysis of a normal individual (top panel) and a patient with X-linked agammaglobulinemia (bottom panel). Blood lymphocytes from a normal individual bind labeled antibody to both the B-cell marker CD19 and the T-cell marker CD3 (see top panel). However, blood lymphocytes from an individual with X-linked agammaglobulinemia such as Bill show only binding to antibodies against the T-cell marker CD3. This indicates an absence of B cells in these patients.

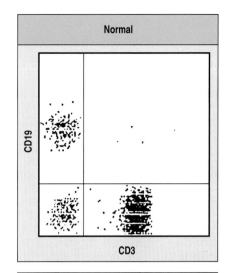

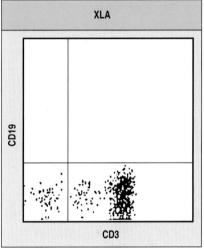

T-cell proliferation indices in response to PHA, concanavalin A, tetanus toxoid and diphtheria toxoid were 162, 104, 10, and 8, respectively (all normal). Serum IgG remained low at 155 mg dl^{-1}, while serum IgA and IgM were undetectable.

Bill was started on a preparation of gamma globulin rendered suitable for intravenous administration. He was given a dose of gamma globulin intravenously so as to maintain his IgG level at 600 mg dl^{-1}. He improved remarkably. The rales at his lung bases disappeared. He continued to perform well in school and ultimately entered medical school. Except for occasional bouts of conjunctivitis or sinusitis, which respond well to oral antibiotic treatment, he remains in good health and leads an active life. He became skilled at inserting a needle into a vein on the back of his hand and he infuses himself with 10 g gamma globulin every weekend.

X-linked agammaglobulinemia.

Males such as Bill with a hereditary inability to make antibodies are subject to recurrent infections. However, the infections are due almost exclusively to common extracellular bacterial pathogens—*Haemophilus influenzae*, *Streptococcus pneumoniae*, *Streptococcus pyogenes*, and *Staphylococcus aureus*. An examination of scores of histories of boys with this defect has established that they have no problems with intracellular infections, such as those caused by the common viral diseases of childhood. T-cell number and function in males with X-linked agammaglobulinemia are normal, and they therefore have normal cell-mediated responses, which are able to terminate viral infections and infections with intracellular bacteria such as those causing tuberculosis.

The bacteria that are the major cause of infection in X-linked agamma-globulinemia are all so-called pyogenic bacteria. Pyogenic means pus-forming, and pus consists largely of neutrophils. The normal host response to pyogenic infections is the production of antibodies that coat the bacteria and fix complement, thereby enhancing rapid uptake of the bacteria into phagocytic cells such as neutrophils and macrophages, which destroy them. Since antibiotics came into use, it has been possible to treat pyogenic infections successfully. However, when they recur frequently, the excessive release of proteolytic enzymes (for example elastase) from the bacteria and from the host phagocytes causes anatomical damage, particularly to the airways of the lung. The bronchi lose their elasticity and become the site of chronic inflammation (this is called bronchiectasis). If affected males do not receive replacement therapy—gamma globulin—to prevent pyogenic infections, they eventually die of chronic lung disease.

Gamma globulin is prepared from human plasma. Plasma is pooled from approximately one thousand or more blood donors and is fractionated at very cold temperatures (–5°C) by adding progressively increasing amounts of ethanol. This method was developed by Professor Edwin J. Cohn at the Harvard Medical School during the Second World War. The five plasma

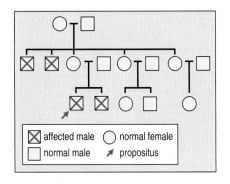

Fig. 2.5 Bill's family tree.

fractions obtained are still called Cohn Fractions I, II, III, IV, and V. Cohn Fraction I is mainly composed of fibrinogen. Cohn Fraction II is almost pure IgG and is called gamma globulin. Cohn Fraction III contains the beta globulins, including IgA and IgM, Fraction IV, the alpha globulins, and Fraction V, albumin. Cohn Fraction II, or gamma globulin, is commercially available as a 16% solution of IgG. During the processing of the plasma some of the gamma globulin aggregates and, for this reason, the 16% solution cannot be given intravenously. Aggregated gamma globulin acts like immune complexes and causes a reaction of shaking chills, fever and low blood pressure when given intravenously. Gamma globulin can be disaggregated with low pH or insoluble proteolytic enzymes. It can then be safely administered intravenously as a 5% solution.

The gene defect in X-linked agammaglobulinemia was identified when the gene was mapped to the long arm of the X chromosome at Xq22 and subsequently cloned. The gene encodes a cytoplasmic tyrosine kinase found in pre-B cells, B cells and neutrophils. It has been named *btk*. *Btk* is not required for neutrophil development and it is not yet known precisely what role it has in the normal maturation of B lymphocytes.

Discussion and questions.

1   *Fig. 2.5 shows Bill's family tree. It can be seen that only males are affected and that the females who carry the defect (Bill's mother and maternal grandmother) are normal. This inheritance pattern is characteristic of an X-linked recessive trait. We do not know whether Bill's aunts are carriers of the defect because neither of them has had an affected male child. Now that the btk gene has been mapped, it is possible in principle to detect carriers by testing for the presence of a mutant btk gene. But there is a much simpler test that was already available at the time of Bill's diagnosis, which is still used routinely. Can you suggest how we could have determined whether Bill's aunts were carriers?*

In every somatic cell of a female, one of the two X chromosomes is inactivated. Which of the X chromosomes is inactivated is a random process so that each is normally active in 50% of the cells on average. However, if the normal X chromosome is inactivated in a pre-B cell of Bill's mother or grandmother, that cell has no normal *btk* product and cannot mature. All of their B cells therefore have the normal X chromosome active (Fig. 2.6). This makes it appear that in the B cells of Bill's mother and grandmother the inactivation of the X chromosome has been non-random. If we have a marker that allows us to distinguish between the two X chromosomes of Bill's aunts we could determine whether their B cells exhibit random or non-random X inactivation. In fact it turns out that one of Bill's aunts is a carrier and the other is not.

2   *Bill was well for the first 10 months of his life. How do you explain this?*

Maternal IgG crossed the placenta into Bill's circulation during fetal life. He was protected passively by the maternal IgG for 10 months.

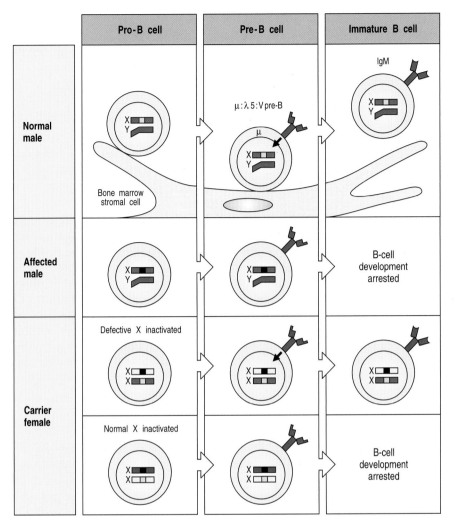

Fig. 2.6 Arrested B-cell development in X-linked agammaglobulinemia is responsible for apparently non-random X inactivation in the B cells of female carriers. In normal individuals, B-cell development proceeds through a stage in which the pre-B-cell receptor consisting of μ:λ5:Vpre-B transduces a signal via Btk, triggering further B-cell development. In males with XLA, no signal can be transduced and, although the pre-B-cell receptor is expressed, the B cells develop no further. In female mammals, including humans, one of the two X chromosomes in each cell is permanently inactivated early in development. Because the choice of which chromosome to inactivate is random, half of the pre-B cells in a carrier female express a wild-type *btk*, and half express the defective gene. None of the B cells that express *btk* from the defective chromosome can develop into mature B cells. Therefore, in the carrier, mature B cells always have the non-defective X chromosome active. This is in sharp contrast to all other cell types, which express the non-defective chromosome in only half of the population. Apparently non-random X chromosome inactivation in a particular cell lineage is a clear indication that the product of the X-linked gene is required for the development of cells of that lineage. It is also sometimes possible to identify the stage at which the gene product is required, by detecting the point in development at which X-chromosome inactivation develops bias. Using this kind of analysis, one can identify carriers of traits like XLA without needing to know the nature of the gene.

---

3 *Patients with immunodeficiency diseases should never be given live viral vaccines! Several male infants with X-linked agammaglobulinemia have been given live oral polio vaccine and have developed paralytic poliomyelitis. What sequence of events led to the development of polio in these boys?*

Live polio vaccines are made from viruses with a disabling mutation in the gene that allows the virus to enter the motor nerve cells and cause paralysis. When a normal infant is given live, attenuated poliovirus orally, the poliovirus establishes a harmless infection in the gut. Within 2 weeks the infant makes IgG and IgA antibodies that neutralize the poliovirus and prevent the infection from spreading, so that as the infected gut cells die the infection is terminated. When male infants with X-linked agammaglobulinemia are given this same vaccine, they are incapable of making any antibodies and the infection can persist. They continue to excrete the poliovirus from their gut. After a time there may be a mutation in the some of the viruses that causes them to re-acquire the ability to enter nerve cells (neurotropism). These so-called revertant viruses disseminate through the bloodstream and infect the neurons in the spinal cord, thus causing paralytic poliomyelitis. Another example of a virus that may disseminate from the gastrointestinal tract in males with X-linked agammaglobulinemia are the echoviruses. These are cleared readily by normal individuals, but without the benefit of either IgA or IgG antibodies against the infecting serotype the virus can disseminate to the central nervous system and cause meningoencephalitis.

4   *Bill has a normal number of lymphocytes in his blood (43% of a normal concentration of 5100 µl⁻¹ white blood cells). Only by phenotyping these lymphocytes do we realize that they are all T cells (CD3⁺) and that he has no B cells (CD19⁺). What tests were performed to establish that his T cells function normally?*

Certain plant lectins, such as phytohemagglutinin and concanavalin A, cause virtually all T cells to divide and are therefore known as nonspecific mitogens. Antigens to which the host has previously been exposed also cause T cells to divide *in vitro*. After 72 hours, exposure either to nonspecific mitogens or to specific antigens, ³H-thymidine (tritiated thymidine) was added to Bill's T-cell cultures. Tritiated thymidine becomes incorporated into the DNA of dividing cells. The stimulation indices (the number of tritium counts in the stimulated cultures divided by the number of counts in similar cultures not exposed to mitogen or antigen) were normal for both mitogens and antigens. Bill's T cells responded to tetanus and diphtheria toxoids because he had been immunized with these inactivated bacterial toxins before the diagnosis was established.

5   *Bill's recurrent infections were due almost exclusively to* Streptococcus *and* Haemophilus *species. These bacteria have a slimy capsule composed primarily of polysaccharide polymers, which protect them from direct attack by phagocytes. Humans make IgG2 antibodies to these polysaccharide polymers. The IgG2 antibodies 'opsonize' the bacteria by fixing complement on their surface, thereby facilitating the rapid uptake of these bacteria by phagocytic cells (Fig. 2.7). What other genetic defect in the immune system might clinically mimic X-linked agammaglobulinemia?*

Hereditary deficiency of complement component C3. The fixation of C3 to the bacterial surface, either by the classical or by the alternative pathways of complement activation, leads to its cleavage into a succession of fragments, two of which (C3b and C3bi) bind to complement receptors on the surface of phagocytic cells and enhance phagocytosis. C3bi binds the most potent complement receptor (CR3), and is the most important opsonizing agent for the ingestion and phagocytosis of encapsulated bacteria (see Case 10).

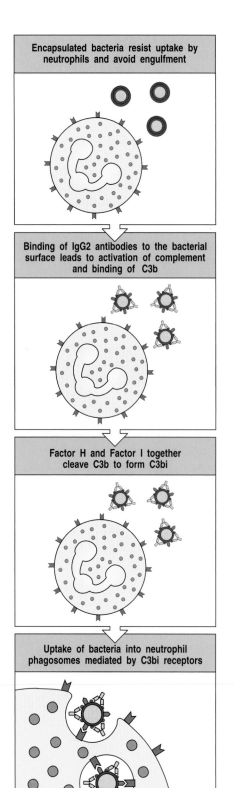

Encapsulated bacteria resist uptake by neutrophils and avoid engulfment

Binding of IgG2 antibodies to the bacterial surface leads to activation of complement and binding of C3b

Factor H and Factor I together cleave C3b to form C3bi

Uptake of bacteria into neutrophil phagosomes mediated by C3bi receptors

Fig. 2.7 Encapsulated bacteria are efficiently engulfed by phagocytes only when they are coated with complement. Encapsulated bacteria resist ingestion by phagocytes unless they are recognized by antibodies that fix complement. IgG2 antibodies are produced against these bacteria in man, and lead to the deposition of complement component C3b on the bacterial surface, where it is cleaved by Factor H and Factor I to produce C3bi, still bound to the bacterial surface. C3bi binds a specific receptor on phagocytes and induces the engulfment and destruction of the C3bi-coated bacterium. Phagocytes also have receptors for C3b, but these are most effective when acting in concert with Fc receptors for IgG1 antibodies, whereas the C3bi receptor is potent enough to act alone, and is the most important receptor for the phagocytosis of pyogenic bacteria.

| 6 | *The doctor noted that Bill had no tonsils though he had never had his tonsils removed surgically. How do you explain this absence of tonsils, an important diagnostic clue in suspecting X-linked agammaglobulinemia?* |

Tonsils are 80–90% B cells.

| 7 | *It was found by trial and error that Bill would stay healthy and have no significant infections if his IgG level were maintained at 600 mg dl⁻¹ of plasma. He was told to take 10 g of gamma globulin every week to maintain that level. How was the dose calculated?* |

The rate at which IgG is catabolized depends on its concentration. In normal individuals, IgG has a half-life of approximately 21 days. In males with X-linked agammaglobulinemia, because the concentration is lower, IgG has a half-life of approximately 28 days. Overall, Bill's optimal level of 600 mg dl^{-1} decreases to about 450 mg dl^{-1} after a week because of catabolism, as well as minor losses in saliva, tears, the gut and other secretions.

IgG, like all plasma proteins, distributes into the extravascular space: half the body IgG is in the blood and the other half is in the extravascular space. A dose of gamma globulin administered intravenously would equilibrate with the extravascular fluid in 24 hours. The dose of 10 g is arrived at as shown in Fig. 2.8.

Bill weighs 75 kg.

The vascular volume is 8% of body weight or 0.08 liters kg^{-1}.

Bill's blood volume is therefore 75 × 0.08 = 6 liters.

Bill's hematocrit (the portion of the blood composed of cells) is 45%.

Therefore his plasma is 55% of his blood volume or 6 × 0.55 = 3.3 liters (3300 ml).

Bill injects himself with 10 g gamma globulin or 10,000 mg. He thereby raises his IgG level by 10,000/3300 or roughly 3 mg ml^{-1} or 300 mg dl^{-1}.

Half the IgG equilibrates into the extravascular pool, so that Bill has really raised his plasma level by only 150 mg dl^{-1}.

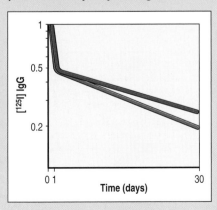

Fig. 2.8 IgG was radiolabeled with ^{125}I and then injected intravenously. After 10 minutes to allow for complete mixing of the radioactive dose in the blood, a plasma sample was obtained and the radioactivity was assayed. The amount of radioactivity at that time point was considered to be 100%. Plasma samples were obtained subsequently at frequent intervals and the percentage of residual radioactivity was determined and plotted on semi-log graph paper. The green line shows the rate of disappearance of IgG in a normal person and the red line the rate in a male with X-linked agammaglobulinemia. From inspection of the curve it can be determined how long it takes for the radioactivity to decrease by 50%, say from 50% to 25%.

CASE 3 | Hyper IgM Immunodeficiency

Failure of immunoglobulin isotype switching.

After exposure to an antigen, the first antibodies to appear are IgM. Later, antibodies of other classes appear: IgG predominates in the serum and extravascular space, while IgA is produced in the gut, and IgE may be secreted at other epithelial surfaces. The different effector functions of these different classes, or isotypes, are summarized in Fig. 3.1. The changes in the isotype of antibody produced in the course of an immune response reflect the occurrence

Functional activity	IgM	IgD	IgG1	IgG2	IgG3	IgG4	IgA	IgE
Neutralization	+	–	++	++	++	++	++	–
Opsonization	–	–	+++	*	++	+	+	–
Sensitization for killing by NK cells	–	–	++	–	++	–	–	–
Sensitization of mast cells	–	–	+	–	+	–	–	+++
Activates complement system	+++	–	++	+	+++	–	+	–

Distribution	IgM	IgD	IgG1	IgG2	IgG3	IgG4	IgA	IgE
Transport across epithelium	+	–	–	–	–	–	+++ (dimer)	–
Transport across placenta	–	–	+++	+	++	+/–	–	–
Diffusion into extravascular sites	+/–	–	+++	+++	+++	+++	++ (monomer)	+
Mean serum level (mg ml^{-1})	1.5	0.04	9	3	1	0.5	2.1	3×10^{-5}

Fig. 3.1 Each human immunoglobulin isotype has specialized functions and a unique distribution. The major effector functions of each isotype (+++) are shaded in dark red, while lesser functions (++) are shown in dark pink, and very minor functions (+) in pale pink. The distributions are similarly marked with actual average levels in serum shown in the bottom row. *IgG2 can act as an opsonin in the presence of Fc receptors of a particular allotype, found in about 50% of Caucasians.

Topics bearing on this case:

Antibody isotypes

Isotype switching

CD40 ligand and isotype switching

Antibody-mediated bacterial killing

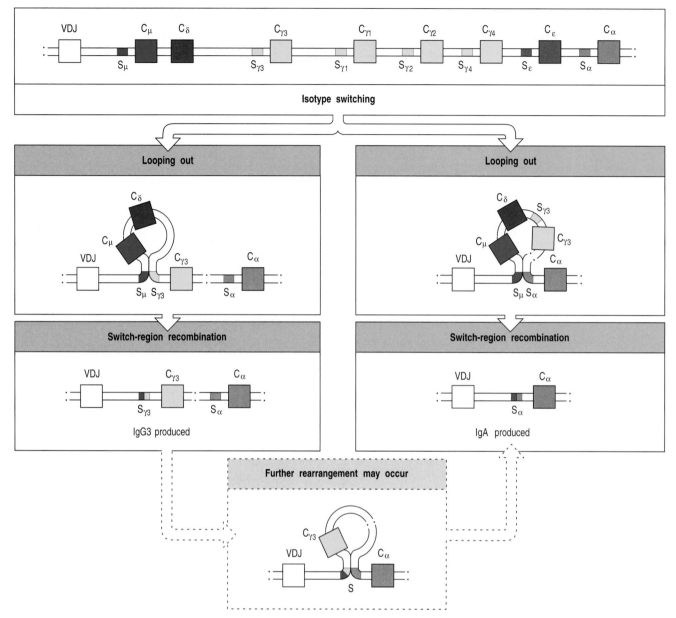

Fig. 3.2 Isotype switching involves recombination between specific signals. Repetitive DNA sequences that guide isotype switching are found upstream of each of the immunoglobulin C-region genes, with the exception of the δ gene. Switching occurs by recombination between these repetitive sequences or switch signals, with deletion of the intervening DNA. The initial switching event takes place from the μ switch region: switching to other isotypes can take place subsequently from the recombinant switch region formed after μ switching. S = switch region.

of isotype switching in the B cells that synthesize antibody, so that the variable region, which determines the specificity of an antibody, becomes associated with the constant regions of different isotypes as the immune response progresses (Fig. 3.2).

Isotype switching is induced by T cells. T cells are also required to initiate B-cell responses to many antigens: the only exceptions are responses triggered by some microbial antigens, or by certain antigens with repeating epitopes. T-cell help is delivered in the context of an antigen-specific interaction with the B cell (Fig. 3.3). This interaction activates the T cell to express the CD40 ligand (CD40L), which then delivers an activating signal to the B cell by binding

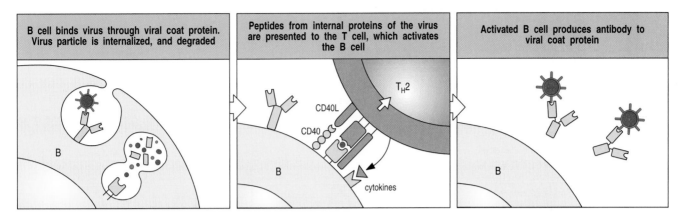

Fig. 3.3 B cells are activated by helper T cells that recognize antigenic peptide bound to class II molecules on their surface. An epitope on a viral coat (spike) protein is recognized by the surface immunoglobulin on a B cell and the virus is internalized and degraded. Peptides derived from viral proteins are returned to the B-cell surface bound to MHC class II molecules, where they are recognized by previously activated helper T cells that activate the B cells to produce antibody against the virus.

CD40. Activated T cells also secrete cytokines, which are required at the initiation of the humoral immune response to drive the proliferation and differentiation of naive B cells, and are later required to induce isotype switching (Fig. 3.4). In humans, isotype switching to IgE synthesis is best understood and is known to require interleukin-4 or interleukin-13, as well as contact with CD40L expressed by an activated T cell.

The gene for the CD40 ligand is on the X chromosome at position Xq26. In males with a defect in the CD40 ligand gene, isotype switching fails to occur; such individuals make only IgM and IgD and cannot switch to IgG, IgA, or IgE synthesis. This defect can be mimicked in mice in which the genes for CD40

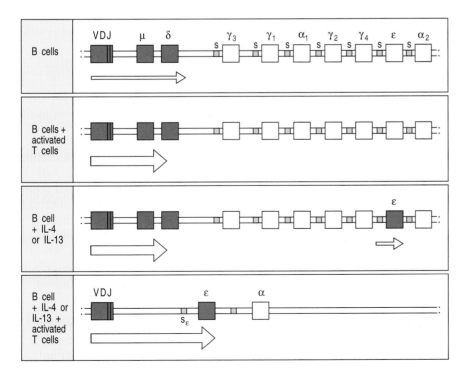

Fig. 3.4 Isotype switching to IgE production by human B cells. Purified human B cells in culture transcribe the μ and δ loci at a low rate, giving rise to surface IgM and IgD. On co-culture with T cells activated with ionomycin and phorbol myristate acetate (PMA), IgM is secreted. The presence of IL-4 or IL-13 stimulates an isotype switch to IgE. Purified B cells cultured alone with these cytokines transcribe the Cε gene at a low rate, but the transcripts originate in the switch region preceding the gene and do not code for protein. On co-culture with activated T cells in addition to IL-4 or IL-13, an isotype switch occurs, mature ε RNA is expressed and IgE synthesis and secretion proceed.

Fig. 3.5 Flow cytometric analysis showing that activated T cells from hyper IgM patients do not express the CD40 ligand. T cells from two patients and one healthy donor were activated *in vitro* with a T-cell mitogen, incubated with soluble CD40 protein and analyzed by flow cytometry (see Fig. 2.3). The results are shown in the top three panels. In the normal individual, there are two populations of cells: one that does not bind CD40 (the peak to the left, with low-intensity fluorescence) and one that does (the peak to the right, with high-intensity fluorescence). The dotted line is the negative control, showing non-specific binding of a fluorescently labeled protein to the same cells. In the hyper IgM patients (middle and right-hand panels), CD40 fluorescence exactly coincides with the non-specific control, showing that there is no specific binding to CD40 by these cells. The bottom panels show that the T cells have indeed been activated by the mitogen for the T cells of both the normal and the two patients have increased expression of the interleukin-2 receptor, CD25, as anticipated after T-cell activation. The dotted line is a negative control of fluorescent goat anti-mouse immunoglobulin.

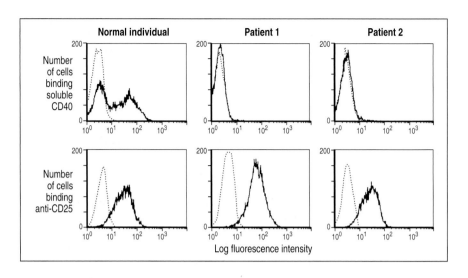

or CD40 ligand have been disrupted by gene targeting: B cells in these animals also fail to undergo isotype switching. The underlying defect in patients with hyper IgM immunodeficiency syndrome can be demonstrated easily by using soluble CD40. Soluble CD40 can be made by engineering the extracellular domain of CD40 onto the constant region of IgG. The soluble protein can then be labeled with a dye. Activated T cells from hyper IgM patients fail to bind fluorescently labeled, soluble CD40 (Fig. 3.5).

CD40 is expressed not only on B cells but also on the surfaces of macrophages, dendritic cells, follicular dendritic cells and mast cells. Macrophages and dendritic cells are professional antigen-presenting cells that can trigger the initial activation and expansion of antigen-specific T cells at the start of an immune response. Recent experiments in CD40L-deficient mice indicate a role for the CD40:CD40L interaction in this early priming event, because in the absence of CD40L the initial activation and expansion of T cells in response to protein antigens is greatly reduced.

The case of Dennis Fawcett: a failure of T-cell help.

Five-year-old boy fails to make antibody to strep infection.

Dennis Fawcett was 5 years old when he was referred to the Children's Hospital with a severe acute infection of the ethmoid sinuses (ethmoiditis). His mother reported that he had had recurrent sinus infections since he was 1 year old. Dennis had pneumonia from an infection with *Pneumocystis carinii* when he was 3 years old. These infections were treated successfully with antibiotics. While he was in the hospital with ethmoiditis, group A β-hemolytic streptococci were cultured from his nose and throat. The physicians caring for Dennis expected that he would have a brisk rise in his white blood cell count as a result of his severe bacterial infection, yet his white blood cell count was 4200 μl^{-1} (normal count 5000–9000 μl^{-1}). 26% of his white blood cells were neutrophils, 56% lymphocytes and 28% monocytes. Thus his neutrophil number was very low, whereas his lymphocyte number was normal and the number of monocytes was elevated.

Seven days after admission to the hospital, during which time he was successfully treated with intravenous antibiotics, his serum was tested for antibodies to streptolysin O, an antigen secreted by streptococci. When no antibodies to the streptococcal antigen were found, his serum immunoglobulins were measured. The IgG level was 25 mg dl^{-1} (normal 600–1500 mg dl^{-1}), IgA was undetectable (normal 150–225 mg dl^{-1}) and his IgM level was elevated at 210 mg dl^{-1} (normal 75–150 mg dl^{-1}). A lymph-node biopsy showed poorly organized structures with an absence of secondary follicles and germinal centers (Fig. 3.6).

Dennis was given a booster injection of diphtheria toxoid, pertussis and tetanus toxoid (DPT) as well as typhoid vaccine. 14 days later, no antibody was detected to tetanus toxoid nor to typhoid O and H antigens. Dennis had red blood cells of group O. People with type O red blood cells make antibodies to the A substance of type A red cells and antibodies to the B substance of type B red cells. This is because bacteria in the intestine have antigens that are closely related to A and B antigens. Dennis' anti-A titer was 1:3200 and his anti-B titer 1:800, both very elevated. His anti-A and anti-B antibodies were of the IgM class only.

His peripheral blood lymphocytes were examined by FACS analysis and normal results were obtained: 11% reacted with an antibody to CD19, a B-cell marker, 87% with anti-CD3, a T-cell marker, and 2% with anti-CD56, a marker for natural killer (NK) cells. However, all of his B cells (CD19$^+$) had surface IgM and IgD and none were found with surface IgG or IgA. His activated T cells did not bind soluble CD40.

Dennis had an older brother and sister. They were both well. There was no family history of unusual susceptibility to infection.

Dennis was treated with intravenous gamma globulin, 600 mg per kg body weight each month, and subsequently remained free of infection.

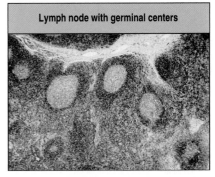

Lymph node from patient with hyper IgM syndrome (no germinal centers)

Lymph node with germinal centers

Fig. 3.6 Photographs of lymph nodes. Bottom panel courtesy of A Perez-Atayde.

Hyper IgM immunodeficiency.

Boys with a hereditary deficiency of the CD40 ligand exhibit consequences of a defect in both humoral and cell-mediated immunity. As we saw in Case 2, defects in antibody synthesis result in susceptibility to the so-called pyogenic infections. These infections are caused by pus-forming (pyogenic) bacteria such as *Haemophilus influenzae, Streptococcus pneumoniae, Streptococcus pyogenes* and *Staphylococcus aureus*, which are resistant to destruction by phagocytic cells unless they are coated (opsonized) with antibody and complement. On the other hand, defects in cellular immunity result in susceptibility to opportunistic infections. Bacteria, viruses, and fungi that normally reside in our bodies and only cause disease when cell-mediated immunity in the host is defective are said to cause opportunistic infections.

Dennis revealed susceptibility to both kinds of infection. His recurrent sinusitis, as we have seen, was caused by *Streptococcus pyogenes*, a pyogenic infection. He also had pneumonia caused by *Pneumocystis carinii*, a protozoan that is ubiquitous and causes opportunistic infections in individuals with a defect in cell-mediated immunity.

Boys with a CD40 ligand deficiency can make IgM in response to T-cell independent antigens but they are unable to make antibodies of any other isotype, and they cannot make antibodies to T-cell dependent antigens, leaving them largely unprotected from many bacteria. They also have a defect in cell-mediated immunity that strongly suggests a role for CD40L in the T-cell mediated activation of macrophages.

The failure of this interaction also explains why affected males are unable to develop leukocytosis (a significant rise in the white count) in the face of severe infections. At times these boys may become profoundly deficient in neutrophils, and this is a very prominent feature of their disease. As a consequence of this neutropenia they develop severe sores and blisters in their mouth and throat, a site normally infested with many bacteria. This defect can be overcome by giving them granulocyte–macrophage cell-stimulating factor (GM-CSF). GM-CSF is secreted by macrophages and to a lesser extent by T cells. The interaction of the CD40 ligand with CD40 on macrophages is required for the secretion of GM-CSF by macrophages, thus explaining why these patients cannot develop a leukocytosis.

Discussion and questions.

---

1 Dennis' B cells expressed IgD as well as IgM on their surface. Why did he not have any difficulty in isotype switching from IgM to IgD?

There is no DNA switch region 5′ to the Cδ gene. A single transcript of VDJCμCδ is alternatively spliced to yield the μ or δ heavy chain (see Fig. 3.4). On the other hand, there are DNA switch regions 5′ to all the other heavy-chain C genes, and isotype switching must occur before functional transcripts of these genes are made. Isotype switching requires recombinase enzymes, which are activated in response to signals from T cells.

2 Normal mice are resistant to Pneumocystis carinii. SCID mice, which have no T or B cells but normal macrophages and monocytes, are susceptible to this infection. In normal mice, Pneumocystis carinii are taken up and destroyed by macrophages. Macrophages express CD40. When SCID mice are reconstituted with normal T cells they acquire resistance to Pneumocystis infection. This can be abrogated by antibodies to the CD40 ligand. What do these experiments tell us about this infection in Dennis?

These experiments tell us that the activation of macrophages against this opportunistic microorganism requires binding of CD40 on the macrophage surface to the CD40 ligand expressed on activated T cells.

3 Why did Dennis make antibodies to blood group A and B antigens but not to tetanus toxoid, typhoid O and H, and streptolysin antigens? Would he have made any antibodies in response to his Streptococcus pyogenes infection?

Hyper IgM patients are deficient in antibody responses to T-cell dependent antigens but can make IgM antibodies to antigens that can stimulate a B-cell response without T-cell help. The blood group antigens are sugar groups that are also found on bacteria in the gut and can activate B cells in the absence of T-cell help. The secreted bacterial toxins, on the other hand, are proteins that cannot elicit a B-cell response in the absence of T cells. Without CD40 ligand, Dennis' T cells would have been unable to activate his B cells to respond to these protein antigens. He would also have been unable to make antibodies to Streptococcus pyogenes, because the antigenic component of the bacterial capsule is a protein.

4   *Most IgM is in the blood and less than 30% of IgM molecules get into the extravascular fluid. On the other hand, over 50% of IgG molecules are in the extravascular space. Furthermore we have 30 to 50 times more IgG molecules than IgM molecules in our body. Why are IgG antibodies more important in protection against pyogenic bacteria?*

The polysaccharide capsules of these pyogenic bacteria are resistant to destruction by phagocytes unless they are opsonized. IgM is not, in itself, an opsonin because there are no receptors on phagocytic cells for the Fc portion of IgM. IgM can promote the phagocytosis of bacteria only by activating complement, leading to the deposition of C3 fragments on the bacterial surface. The C3 is recognized by complement receptors on the phagocytic cells. IgG, however, is more efficient than IgM in promoting the phagocytosis of most bacteria. In addition, there is a range of Fc receptors for IgG isotypes on phagocytes, and IgG1, IgG2 and IgG3 antibodies all promote complement activation on bacterial surfaces. This means that bacteria coated with IgG stimulate phagocytosis through two different classes of receptors, Fc and C3. This results in much more efficient phagocytosis than stimulation through a single class of receptor.

5   *Newborns have difficulty in transcription of the CD40 ligand gene. Does this help to explain the susceptibility of newborns to pyogenic infections? Cyclosporin A, a drug widely used for immunosuppression in graft recipients, also inhibits the transcription of the CD40 ligand gene. What does this imply for patients taking this drug?*

Both newborns and people taking immunosuppressive drugs such as cyclosporin A exhibit increased susceptibility to both pyogenic and opportunistic infections.

A second reason for the increased susceptibility of newborns to some pyogenic infections is the immaturity of many of their B cells. Newborns are normally protected until their lymphocytes mature by pre-existing maternal IgG.

Cyclosporin A also inhibits transcription of the IL-2 gene, thereby preventing the expansion of T-cell clones activated by antigen. This means that all T-cell mediated immune responses, including cytotoxic T-cell responses, are suppressed by cyclosporin A.

6   *Dennis has an X-linked immunodeficiency and yet there is no informative family history. Is he a new mutation or is his mother a carrier of the defect in the CD40 ligand gene? Can we test his sister to determine whether she is a carrier? In the case of X-linked agammaglobulinemia we learned of the usefulness of examining random or non-random X-chromosome inactivation to answer these questions about the carrier state. Would it be useful to examine the T cells of Dennis' mother and sister for non-random X inactivation?*

In a word, no! In the case of X-linked agammaglobulinemia we saw that the gene defect in a tyrosine kinase, Btk, was critical for the normal maturation of B cells. Thus pro-B cells with the X chromosome carrying the *btk* mutation did not mature; only the pro-B cells with the normal X chromosome matured.

This appeared as non-random X-chromosome inactivation in the B cells of the carriers. The CD40 ligand is not needed for the normal maturation of T cells. Its expression is induced in mature T cells by activating them. Dennis' mother and sister exhibit random X-chromosome inactivation in their T cells and this test is of no help in determining the carrier state in the hyper IgM immunodeficiency syndrome. The defect in Dennis' CD40L gene may be either a deletion or a missense mutation, and would need to be analyzed to find a genetic marker that could then be looked for in his mother and sister.

CASE 4 MHC Class I Deficiency

A failure of antigen processing.

The class I molecules encoded by the major histocompatibility complex (MHC) are expressed to a greater or lesser extent on the surface of all the cells of the body except the red blood cells. MHC class I molecules bind peptides derived from proteins synthesized in the cytoplasm, and carry them to the cell surface, where they form a complex of peptide and MHC molecule on the cell surface. This complex can then be recognized by antigen-specific CD8 T cells. T cells as a class recognize only peptides presented to them as a complex with an MHC molecule; the T-cell receptors of CD8 T cells recognize only peptides presented by MHC class I molecules, while those of CD4 T cells recognize only peptides presented by MHC class II molecules (see Case 18).

Topics bearing on this case:

Genetic organization of the MHC

Processing of intracytoplasmic protein antigens

Transport of MHC class I molecules to the cell surface

FACS analysis

Effector CD8 T-cell function in virus infection

MHC class I molecules and CD8 T-cell intrathymic maturation

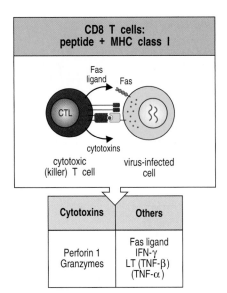

CD8 T cells: peptide + MHC class I	
Cytotoxins	**Others**
Perforin 1 Granzymes	Fas ligand IFN-γ LT (TNF-β) (TNF-α)

Fig. 4.1 Cytotoxic CD8 T cells recognize and kill virus-infected cells. CD8 T cells recognize virally derived peptides presented by MHC class I antigens at the surface of virus-infected cells. They kill target cells by releasing cytotoxins and cytotoxic cytokines and by binding to Fas on the target cell.

MHC class I molecules are involved principally in immune reactions against virus infections. Cytotoxic CD8 T cells specific for viral antigens terminate viral infections by recognizing viral peptides carried by MHC class I molecules on the surface of virus-infected cells, and killing these cells (Fig. 4.1). They release the cytotoxic proteins perforin and granzymes, as well as the cytotoxic cytokines tumor necrosis factors TNF-α and TNF-β. In addition, cytotoxic T cells express the Fas ligand, which engages the cell-surface molecule Fas on target cells. Both processes induce the target cells to undergo programmed cell death (apoptosis).

In humans, the class I and class II MHC molecules are known as the HLA antigens and together they determine the tissue type of an individual. MHC class I molecules are particularly abundant on T and B lymphocytes, and also on macrophages and other cells of the monocyte lineage as well as on neutrophils. Other cells express smaller amounts. Each individual expresses three principal types of class I molecule—HLA-A, HLA-B, and HLA-C. These are heterodimeric glycoproteins, composed of an α chain and a β chain, the latter known also as β_2-microglobulin (Fig. 4.2).

The genes encoding the α chains of the human MHC class I molecules are located close together in the major histocompatibility complex on the short arm of chromosome 6 (Fig. 4.3). In humans, the gene encoding β_2-microglobulin, the polypeptide chain common to all class I molecules, is located not in the MHC but on the long arm of chromosome 15. The genes encoding the α chains are highly polymorphic, and so there are numerous variants of HLA-A, HLA-B, and HLA-C within the population.

Viral proteins, like cellular proteins, are made in the cytoplasm of the infected cell, and some are soon degraded into peptide fragments by large enzyme complexes called proteasomes. The peptides are then transported from the cytosol into the endoplasmic reticulum by a complex of two transporter proteins called TAP-1 and TAP-2, which is located in the endoplasmic reticulum membrane (Fig. 4.4). The genes encoding TAP-1 and TAP-2 are also located in the MHC, in the region containing the class II genes (see Fig. 4.3).

The endoplasmic reticulum contains MHC class I molecules, which enter as separate α and β chains as soon as they have been synthesized and are retained there. After the antigenic peptides enter the endoplasmic reticulum they are loaded onto the complex of α chain and β_2-microglobulin. The MHC class I:peptide complex is then released and transported onwards to the cell surface (Fig. 4.5).

This case describes a rare inherited immune deficiency accompanied by the absence of MHC class I molecules on the patients' cells.

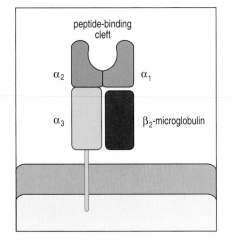

Fig. 4.2 A schematic representation of a human MHC class I molecule. It is a heterodimeric glycoprotein, composed of one transmembrane α chain bound noncovalently to β_2-microglobulin. The α chain is folded into three protein domains, two of which form a cleft into which the peptide antigen binds.

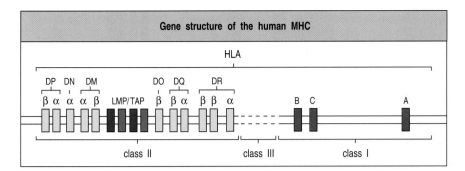

Gene structure of the human MHC

Fig. 4.3 The organization of the major histocompatibility complex (MHC) on chromosome 6 in humans. There are separate regions of class I and class II genes. The class I genes are called HLA-A, HLA-B and HLA-C. The gene for β_2-microglobulin is located on chromosome 15. The genes for the TAP-1:TAP-2 transporter are located in the class II region of the MHC.

The children of Sergei and Natasha Islayev: the consequences of a small flaw in the MHC.

Tatiana Islayev was 17 years old when first seen at the Children's Hospital. She had severe bronchiectasis (dilatation of the bronchi from repeated infections) and a persistent cough that produced yellow-green sputum. Tatiana had been chronically ill from the age of 4, when she started to get repeated infections of the sinuses, middle ears, and lungs, apparently due to a variety of respiratory viruses. The bacteria *Haemophilus influenzae* and *Streptococcus pneumoniae* could be cultured from her sputum and she had been prescribed frequent antibiotic treatment to control her persistent fevers and cough. Her brother Alexander, aged 7 years old, also suffered from chronic respiratory infections. Like his sister, he had begun to suffer severe repeated viral infections of the upper and lower respiratory tracts at an early age. He also had severe bronchiectasis and *H. influenzae* could be cultured from his sputum.

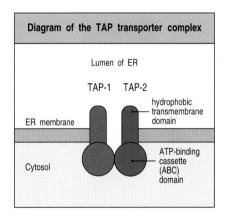

Diagram of the TAP transporter complex

Fig. 4.4 The TAP-1 and TAP-2 transporter proteins form a heterodimer in the endoplasmic reticulum membrane.

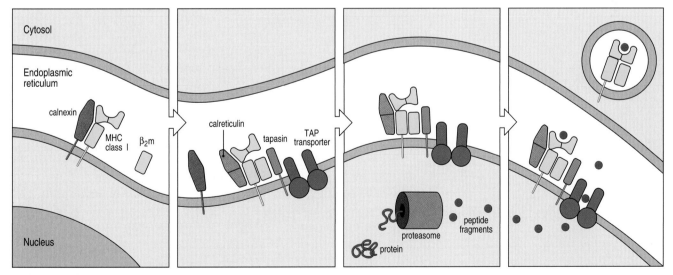

Fig. 4.5 MHC class I molecules do not leave the endoplasmic reticulum unless they bind peptides. MHC class I α chains assemble in the endoplasmic reticulum with a membrane-bound protein, calnexin. When this complex binds β_2-microglobulin (β_2m) it is released from calnexin, and the partly folded MHC class I molecule then binds to the TAP-1 subunit of the TAP transporter by interacting with one molecule of the TAP-associated protein, tapasin. It is retained within the endoplasmic reticulum until released by the binding of a peptide, which completes the folding of the MHC class I molecule. Peptides generated by the degradation of proteins in the cytoplasm are transported into the lumen of the endoplasmic reticulum by the TAP transporter. Once peptide has bound to the MHC molecule, the peptide:MHC complex is transported through the Golgi complex to the cell surface.

Brother and sister with symptoms of severe chronic respiratory infection.

Other siblings normal.

Owing to the chronic illness of Tatiana and Alexander, the Islayevs had emigrated recently from Russia to the United States, where they hoped to get better medical treatment. They had three other children, aged 5, 10, and 13 years old, when they came to America, who were all healthy and showed no increased susceptibility to infection. As infants in Moscow, both Tatiana and Alexander had received routine immunizations with oral poliovirus as well as diphtheria, pertussis and tetanus (DPT) vaccinations. They had also been given BCG as newborn babies for protection against tuberculosis, and had tolerated all these immunizations well.

When they were examined, Tatiana and Alexander both had elevated levels of IgG, at >1500 mg dl^{-1} (normal levels 600–1400 mg dl^{-1}). They had white blood cell counts of 7000 and 6600 cells μl^{-1} respectively. Of these white cells, 25% (1750 and 1650 μl^{-1} respectively) were lymphocytes. Ten per cent of the lymphocytes reacted with an antibody to B cells (anti-CD19) (a normal result) and 4% with an antibody to natural killer (NK) cells (anti-CD16) (normal). The remainder of the lymphocytes reacted with an anti-T-cell antibody (anti-CD3). Over 90% of the T cells were CD4, while 10% were CD8. This represents a profound deficiency of CD8 T cells. Blood tests on their siblings and parents showed no deficiency of CD8 T cells. Both Tatiana and Alexander had normal neutrophil function and complement titers.

There was thus no evidence of a deficiency of humoral immunity. Furthermore, their cell-mediated immunity also seemed normal when tested by delayed hypersensitivity skin tests to tuberculin and antigen from *Candida*, a fungal component of the normal body flora (see Case 8); they developed the normal delayed-type hypersensitivity response of a hard, raised, red swelling some 50 mm in diameter at the site of intradermal injection of these antigens. Both children were found to have high titers of antibodies to herpes virus and cytomegalovirus as well as to mumps, chickenpox and measles viruses. When asked, the parents recalled that the children had been immunized against influenza several times, and antibodies against five different strains of influenza were found. However the anti-influenza antibodies were present in very low titers. They also had low titers of antibody against the Epstein–Barr virus.

When white blood cells from all family members were typed for HLA antigens by standard typing procedures, no MHC class I molecules at all could be found on Tatiana's and Alexander's cells. When their blood cells were examined using the more sensitive technique of FACS analysis (see Fig. 2.3), it was found that Tatiana and Alexander expressed very small amounts of MHC class I molecules, less than 1% of the amount expressed on the cells of their father (Fig. 4.6). They also expressed MHC class II molecules normally.

The HLA typing revealed that the mother and father shared an MHC haplotype (HLA-A3, -B63, HLA-DR4, -DQ3) and that Tatiana and Alexander had inherited this shared haplotype from both their parents (Fig. 4.7). They were therefore homozygous for the MHC region. The other children were heterozygous. It was thus concluded that the two children's susceptibility to respiratory infections was linked to the MHC locus, for which only they were homozygous.

To try to determine the underlying defect, B cells from Tatiana and Alexander were established as cell lines in culture by transformation with Epstein–Barr virus. The transformed B cells were examined for the presence of messenger RNA for MHC class I molecules: normal levels were found. This eliminated the possibility that they had a structural or regulatory defect in genes encoding the α chain of the MHC class I molecules. Because the gene for β$_2$-microglobulin maps to chromosome 15, it was highly unlikely that their MHC-linked condition resulted from a defect in that gene. Only one possibility remained to be explored—a mutation in the *TAP-1* or *TAP-2* genes. When the DNA sequences of these genes were determined, both Tatiana and Alexander were found to have the same nonsense mutation in their *TAP-2* genes. Their parents were found to be heterozygous for this mutation.

Fig. 4.6 Fluorescent antibody typing for HLA antigens. The peripheral blood lymphocytes of Tatiana and her father were typed by a fluorescent antibody to the MHC class I molecule HLA-A3. Tatiana's lymphocytes express about 1% of the HLA-A3 expressed on the lymphocytes of her father.

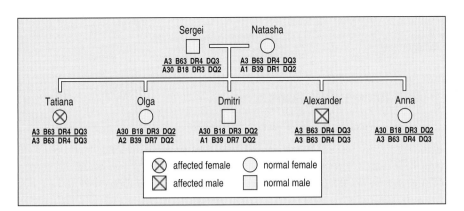

Fig. 4.7 Inheritance of MHC haplotypes in the Islayev family.

MHC class I deficiency.

A recessively inherited immunodeficiency was first suggested by the similarity of Tatiana's and Alexander's condition and its complete absence from other members of the family. The Islayev children are so far unique in the medical literature as they are the only reported cases of MHC class I deficiency for which the molecular basis has been determined.

The Islayev children present us with a unique example of a clinical phenotype resulting from MHC class I deficiency. This phenotype has some unexpected features. One would not have expected their apparently normal capacity to fight some types of viral infections successfully, despite their profound deficiency in CD8 T cells and their inability to present viral antigens to CD8 T cells because of the absence of MHC class I antigens. The high levels of antibodies to chickenpox, measles and mumps viruses in their blood showed that they have been exposed to and had successfully overcome these infections.

They have sustained innumerable respiratory viral infections, however. And their poor antibody responses to a variety of influenza strains show that they may have problems responding to respiratory viruses in general. It may be that some viruses are better able than others to stimulate an increased expression of MHC class I molecules on their cells. Tatiana and Alexander express MHC class I molecules at very low levels and when their isolated B cells were loaded with an antigenic peptide from influenza virus, this stimulated a small increase in the number of MHC class I molecules on these cells (Fig. 4.8). It is therefore possible that other virus infections that they sustained, such as chickenpox, were able to induce sufficient expression of MHC class I molecules to terminate the infection properly.

The repeated respiratory infections have caused anatomic damage to their airways, resulting in the bronchiectasis. The abundant *Haemophilus* and pneumococci in their sputum is characteristic of patients with bronchiectasis, and in their case is not due directly to any deficiency of immunity against these capsulated bacteria (compare with Case 1).

Their profound deficiency in CD8 T cells is a direct consequence of the lack of MHC class I molecules on the cells of the thymus, the organ in which all T cells bearing αβ T-cell receptors mature. An interaction with MHC class I molecules on thymic epithelial cells is crucial for the intrathymic maturation of such T cells as CD8 T cells.

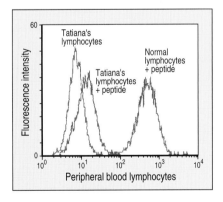

Fig. 4.8 Increased expression of MHC class I molecules in response to virus infection. Peripheral blood lymphocytes from Tatiana were examined with a fluorescent antibody to HLA-A3 before and after loading the cells with an antigenic peptide from influenza virus. There is a small increase in the number of MHC class I molecules expressed on the lymphocytes to approximately 5% of the amount expressed by normal cells loaded with the same peptide.

This family reveals that we have much to learn about the role of MHC class I molecules in protection against intracellular pathogens.

Discussion and questions.

1 *Why was the number of CD8 T cells in Tatiana and Alexander decreased despite normal levels of CD4 T cells?*

The intrathymic maturation of CD8 T cells depends on the expression of MHC class I molecules on thymic epithelial cells. Conversely, maturation as CD4 T cells requires interactions with the MHC class II molecules also present on thymic epithelium. As Tatiana and Alexander do not lack MHC class II molecules, their CD4 T cells are normal and they have normal humoral immunity.

2 *When the T-cell antigen receptors on their CD8 T cells were examined, it turned out that they were all γδ receptors and none were αβ receptors. How do you explain this?*

The maturation of CD8 T cells bearing γδ chains occurs after these cells emigrate from the thymus and is independent of MHC class I expression, whereas the maturation of αβ CD8 T cells occurs in the thymus and is dependent on the expression of MHC class I molecules.

3 *Tatiana and Alexander had normal delayed-type hypersensitivity responses to tuberculin and Candida. Is this surprising in view of their deficiency in CD8 T cells?*

No. Delayed-type hypersensitivity reactions are provoked by antigen-specific CD4 T cells (see Case 8).

4 *Why did Tatiana and Alexander have high levels of serum IgG?*

The factors that help B cells to mature and secrete immunoglobulins are derived from activated CD4 T cells. These cells were normal in Tatiana and Alexander. Factors that suppress B-cell responses are secreted by CD8 T cells, of which the children had very few. They were therefore not very efficient at terminating B-cell mediated humoral immune reactions and tended to over-produce antibody. In Case 18, we shall see the opposite phenomenon in patients with MHC class II deficiency, who have a deficiency of CD4 T cells and very low levels of serum immunoglobulins.

CASE 5 | X-linked Severe Combined Immunodeficiency

The maturation of T lymphocytes.

Without T cells life cannot be sustained. In Case 2 we learned that an absence of B cells was compatible with a normal life style so long as infusions of immunoglobulin G were maintained. When children are born without T cells, they appear normal for the first few weeks or months. Then they begin to acquire opportunistic infections and die while still in infancy. An absence of functional T cells causes severe combined immunodeficiency (SCID). It is severe because it is fatal, and combined because, in humans, B cells cannot function without help from T cells, so that even if the B cells are not directly affected by the defect, both humoral and cell-mediated immunity are lost. Unlike X-linked agammaglobulinemia, which results from a monogenic defect, SCID is a single phenotype that can result from any one of several different genetic defects. The incidence of SCID is three times greater in males than in females and this male:female ratio of 3:1 is due to the fact that the most common form of SCID is X-linked. Approximately 55% of cases of SCID have the X-linked form of the disease.

T-cell precursors migrate to the thymus to mature, at first from the yolk sac of the embryo, and subsequently from the fetal liver and bone marrow (Fig. 5.1). The rudimentary thymus is an epithelial anlage derived from the third and fourth pharyngeal pouches. By 6 weeks of human gestation, the invasion of precursor T cells, and of dendritic and macrophage cells, has transformed the gland into a primary lymphoid organ (Fig. 5.2). T-cell precursors undergo rapid maturation in the thymus gland (Fig. 5.3), which becomes the site of the greatest mitotic activity in the developing fetus. By 20 weeks of gestation mature T cells start to emigrate from the thymus to the secondary lymphoid organs. In all common forms of SCID the thymus fails to become a primary lymphoid organ. A small and dysplastic thymus, as revealed by biopsy, used to be the confirming diagnostic indicator of SCID.

Topics bearing on this case:

Defects in T-cell function result in severe combined immunodeficiency

The development of T cells in the thymus

Testing T-cell responses with polyclonal mitogens

Signal transduction through the T-cell receptor

Interleukin-2 receptor

Fig. 5.1 T-cell precursors migrate to the thymus to mature. T cells derive from bone marrow stem cells, whose progeny migrate from the bone marrow to the thymus (left panel), where the development of the T cell takes place. Mature T cells leave the thymus and recirculate from the bloodstream through secondary lymphoid tissues (right panel) such as lymph nodes, spleen or Peyer's patches, where they may encounter antigen.

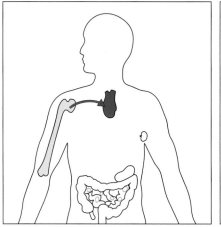 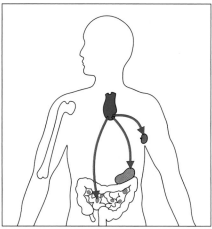

The genetic defect responsible for the X-linked form of SCID has been mapped to the long arm of the X chromosome at Xq11. From this region the gene encoding the γ chain of the interleukin-2 receptor was cloned, and found to be mutated in X-linked SCID.

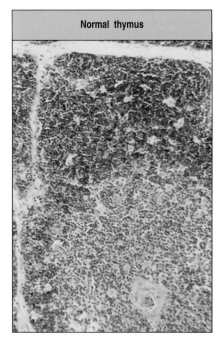

Normal thymus

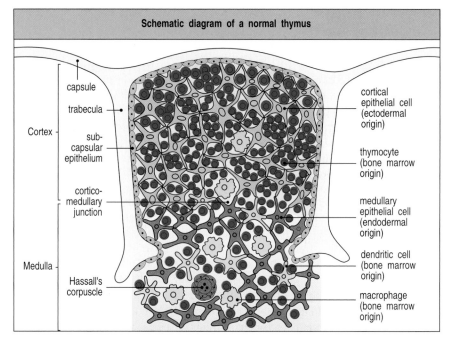

Schematic diagram of a normal thymus

capsule
trabecula
Cortex
sub-capsular epithelium
cortico-medullary junction
Medulla
Hassall's corpuscle

cortical epithelial cell (ectodermal origin)
thymocyte (bone marrow origin)
medullary epithelial cell (endodermal origin)
dendritic cell (bone marrow origin)
macrophage (bone marrow origin)

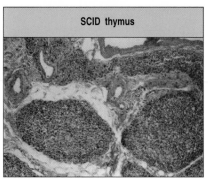

SCID thymus

Fig. 5.2 The cellular organization of the thymus. The thymus, which lies in the midline of the body above the heart, is made up from several lobules, each of which contains discrete cortical (outer) and medullary (central) regions. The cortex consists of immature thymocytes (dark blue), branched cortical epithelial cells (pale blue), with which the immature cortical thymocytes are closely associated, and scattered macrophages (yellow) involved in clearing apoptotic thymocytes. The medulla consists of mature thymocytes (dark blue), and medullary epithelial cells (orange), along with macrophages (yellow) and dendritic cells (yellow) of bone marrow origin. Hassall's corpuscles found in the human thymus are probably also sites of cell destruction. The thymocytes in the outer cortical cell layer are proliferating immature cells, while the deeper cortical thymocytes are mainly cells undergoing thymic selection. The upper photograph shows the equivalent section of a human thymus, stained with hematoxylin and eosin. The lower photograph shows a SCID thymus. There is no corticomedullary differentiation, very low thymocytis and no Hassall's corpuscles. Photograph courtesy of C J Howe (upper) and A Perez-Atayde (lower).

Fig. 5.3 Changes in cell-surface molecules allow thymocyte populations at different stages of maturation to be distinguished. The most important cell-surface molecules for identifying thymocyte subpopulations have been CD4, CD8, and T-cell receptor complex molecules (CD3, and α and β chains). The earliest cell population in the thymus does not express any of these. As these cells do not express CD4 or CD8, they are called 'double-negative' thymocytes. (The γ:δ T cells found in the thymus also lack CD4 or CD8 but these are a minor population.) Maturation of α:β T cells occurs through stages where both CD4 and CD8 are expressed by the same cell, along with the pre-T-cell receptor (pTα:β) and later low levels of the T-cell receptor (α:β) itself. These α:β cells are known as 'double-positive' thymocytes. Most thymocytes (~97%) die within the thymus after becoming small double-positive cells. Those whose receptors bind self MHC molecules lose expression of either CD4 or CD8 and increase the level of expression of the T-cell receptor. The outcome of this process is the 'single-positive' thymocytes, which, after maturation, are exported from the thymus as mature single-positive T cells.

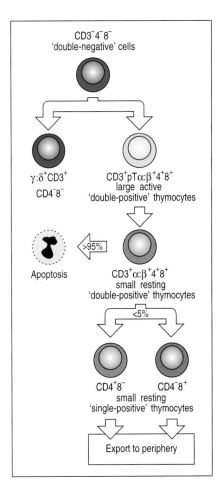

The case of Martin Causubon: without T cells life cannot be sustained.

Mr and Mrs Causubon had a normal daughter 3 years after they were married. Two years later they had a son and named him Martin. He weighed 3.5 kg at birth and appeared to be perfectly normal. At 3 months of age, Martin developed a runny nose and a persistent dry cough. One month later he had a middle ear infection (otitis media) and his pediatrician treated him with amoxicillin. At 5 months of age Martin had a recurrence of otitis media. His cough persisted and a radiological examination of his chest revealed the presence of pneumonia in both lungs. He was treated with another antibiotic, clarithromycin. Mrs Causubon noticed that Martin had thrush (*Candida* spp.) in his mouth (Fig. 5.4) and an angry red rash in the diaper area. He was not gaining weight; he had been in the 50th percentile for weight at age 4 months but by 6 months he had fallen to the 15th percentile. His pediatrician had given him oral polio vaccine at ages 4 and 5 months and, at the same time, diphtheria–pertussis–tetanus (DPT) shots.

Martin's pediatrician referred him to the Children's Hospital for further studies. On admission to the hospital, he was found to be an irritable male infant with tachypnea (fast breathing). A red rash was noted in the diaper area as well as white flecks of thrush on his tongue and buccal mucosa. His tonsils were very small. He had a clear discharge from his nose and cultures of his nasal fluid grew *Pseudomonas aeruginosa*. Coarse, harsh breath sounds were heard in both lungs. His liver was slightly enlarged.

Martin's white blood count was 4800 cells μl^{-1} (normal 5000–10,000 cells μl^{-1}) and his absolute lymphocyte count was 760 cells μl^{-1} (normal 3000 lymphocytes μl^{-1}). None of his lymphocytes reacted with anti-CD3 and it was concluded that he had no T cells. Ninety-nine percent of his lymphocytes bound antibody against the B-cell molecule CD20 and 1% were natural killer cells reacting with anti-CD16. His serum contained IgG at a concentration of 30 mg dl^{-1}, IgA at 27 mg dl^{-1}, and IgM at 42 mg dl^{-1} (IgG levels are normally 400 mg dl^{-1}; the IgA and IgM levels were at the low end of the normal range for Martin's age). His blood mononuclear cells were completely unresponsive to phytohemagglutinin (PHA), concanavalin A (ConA) and pokeweed mitogen (PWM) (Fig. 5.5), as well as to specific antigens to which he had been previously exposed by immunization or infection—tetanus and diphtheria toxoids, and *Candida* antigen. His red cells contained normal amounts of adenosine

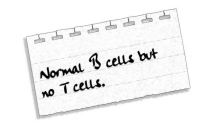

Normal B cells but no T cells.

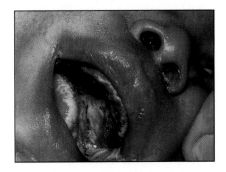

Fig. 5.4 An infant with SCID suffering from *Candida albicans* in the mouth.

Lack of IL-2γ on B cells indicates X-linked SCID.

No HLA-identical sibling. Mother agrees to donate bone marrow.

Engraftment is successful.

deaminase and purine nucleoside phosphorylase. His B lymphocytes did not react with an antibody to the γ chain of the interleukin-2 receptor (IL-2γ). Cultures of sputum for bacteria and viruses revealed the abundant presence of respiratory syncytial virus (RSV).

At this point a blood sample was obtained from Martin's mother in order to examine her T cells for random inactivation of the X chromosome (a diagnostic test explained in Case 2). It was found that her T cells exhibited complete non-random X-chromosome inactivation. HLA typing showed that Martin's sister had no matching HLA alleles. His parents, as expected, each shared one HLA haplotype with Martin.

Martin was treated with intravenous gamma globulin at a dose of 2 g kg^{-1} body weight and his serum IgG level was maintained at 600 mg dl^{-1} by subsequent IgG infusions. He was given aerosolized ribavirin to control his RSV infection and trimethoprim-sulfamethoxazole intravenously for prophylaxis against *Pneumocystis carinii*. Without any further preparation, Martin was given 500×10^6 bone marrow cells from his mother. The bone marrow donation from his mother was depleted of mature T cells by treating with a monoclonal antibody to T cells together with complement. After the transplant of the maternal bone marrow cells Martin was given cyclosporin and prednisone, to suppress any graft-versus-host disease. Sixty days after receiving the maternal bone marrow, Martin's blood contained 1000 maternal CD3$^+$ T cells μl^{-1}, which responded to PHA. His immune system was slowly reconstituted over the ensuing 3 months.

Severe combined immunodeficiency.

Severe combined immunodeficiency, or SCID, presents the physician with a medical emergency. Unless there is a known family history, which provides the opportunity to take corrective therapeutic measures before the onset of infections, children with SCID come to medical attention only after they have been infected with a serious opportunistic infection. As these infants die rapidly from such infections, even when treated adequately with anti-biotics or anti-viral agents, measures must be taken quickly to reconstitute

Fig. 5.5 Polyclonal mitogens, many of plant origin, stimulate lymphocyte proliferation in tissue culture. Many of these mitogens are used to test the ability of lymphocytes in human peripheral blood to proliferate.

Mitogen	Abbreviation	Source	Responding cells
Phytohemagglutinin	PHA	*Phaseolus vulgaris* (Red kidney beans)	T cells (human)
Concanavalin A	ConA	*Canavalia ensiformis* (Jack bean)	T cells
Pokeweed mitogen	PWM	*Phytolacca americana* (pokeweed)	T and B cells
Lipopolysaccharide	LPS	*Escherichia coli*	B cells (mouse)

their immune system. In most cases of SCID, the first symptoms are those of thrush in the mouth and diaper area. A persistent cough usually betrays infection with *Pneumocystis carinii*. The third most common symptom of SCID is intractable diarrhea, usually due to enteropathic coliform bacilli.

As previously mentioned, SCID has many known genetic causes. The autosomal recessive form of SCID is most commonly caused by mutations in the purine degradation enzyme adenosine deaminase (ADA) (see Case 6), and, more rarely, by mutations in another such enzyme, purine nucleoside phosphorylase (PNP). Defects in these enzymes lead to an accumulation of nucleotide substrates that are highly toxic to developing T cells, and to lesser extent also developing B cells. The X-linked form of SCID differs from these autosomal recessive forms in that males with this form of the disease have normal numbers of B cells, but they fail to function in the absence of T cells.

Other cases of autosomal recessive SCID resemble the phenotype of X-linked SCID and have been ascribed to defects in the molecule JAK-3, which transduces the signal from the IL-2 receptor and several other interleukin receptors as well. SCID can also be caused by defects in the γ and ε components of the T-cell receptor, or molecules that transduce signals from the T-cell receptor such as the tyrosine kinase ZAP-70, and the DNA-binding protein NF-AT, as well as by mutations in the interleukin-2 gene itself (Fig. 5.6). In these cases, B cells are normal and at least some T cells are present, but they fail to activate in an adaptive immune response, so that a combined immunodeficiency is seen.

The discovery that mutations in the interleukin-2γ chain (IL-2Rγ) caused X-linked SCID in humans seemed contradictory to the finding that 'knocking-out' the IL-2 gene or genes for other components of the IL-2 receptor did not cause SCID in mice (Fig. 5.7).

This apparent contradiction led to a search for the IL-2γ chain in other interleukin receptors. It was found that this chain also forms part of the IL-4, IL-7, IL-9 and IL-15 receptors, and it was renamed the gamma common (γ_c) chain. JAK-3 transduces the signal from all these receptors by binding the γ_c chain.

Most infants with SCID can be rescued by a successful bone marrow transplant. Continued gamma-globulin therapy is usually necessary but with this, and successfully engrafted T cells, SCID infants survive to lead a relatively normal life. Gene therapy has also been tried successfully in some patients with ADA deficiency.

Discussion and questions.

1   *Martin was suspected to have X-linked SCID because he had normal numbers of B cells and a total absence of T cells in his blood. The diagnosis was further confirmed on finding no IL-2Rγ on the B cells in his blood. What final confirmation proved that he had X-linked SCID?*

His mother had non-random inactivation of the X chromosome in her T cells, thereby demonstrating that she carried an X-linked gene required for the normal maturation of T cells.

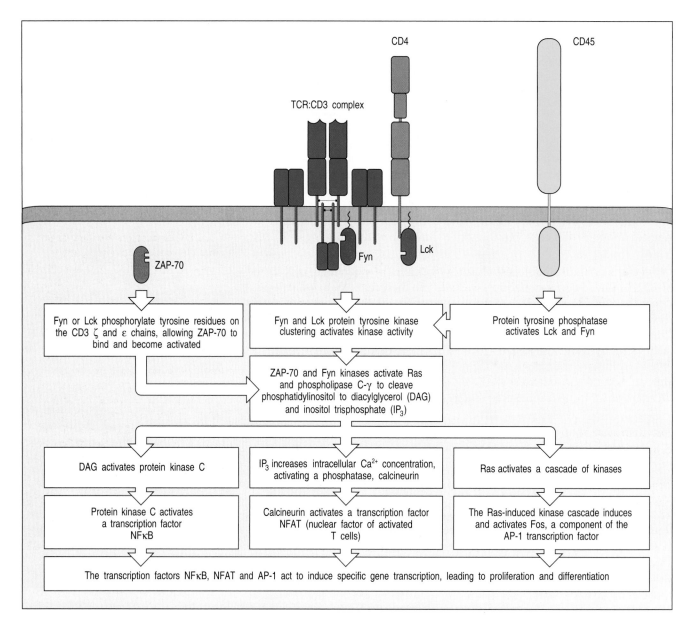

Fig. 5.6 Binding of antigen to the T-cell receptor initiates a series of biochemical changes within the T cell. It is thought that ligand binding to the T-cell receptor and the co-receptor CD4 (in this example) brings together CD4, the T-cell receptor:CD3 complex, and CD45, thus allowing the CD45 tyrosine phosphatase to remove inhibitory phosphate groups and thereby activate the Lck and Fyn tyrosine kinases associated with CD4 and the T-cell receptor:CD3 complex. One effect of tyrosine kinase activation is the activation of the cytosolic tyrosine kinase ZAP-70. Three important signaling pathways result. Two are initiated by the activation of phospholipase C-γ, which cleaves membrane phosphatidylinositol bisphosphate (PIP₂) to yield the second messengers diacylglycerol (DAG) and inositol trisphosphate (IP₃). DAG activates the serine–thronine protein kinase C, which results in the activation of the transcription factor NFκB. Meanwhile, IP₃ acts to release calcium ions (Ca²⁺) from intracellular stores. In addition, a calcium-specific ion channel is opened in the T-cell membrane to allow the influx of Ca²⁺ from extracellular sources. The elevated concentration of Ca²⁺ activates a cytoplasmic phosphatase, calcineurin, which activates one part of the transcription factor NFAT. However, full NFAT activity also requires a member of the AP-1 family of transcription factors; these are dimers of members of the Fos and Jun families of transcription regulators. A third signaling pathway involves activation of the G-protein Ras, and the subsequent activation of a cascade of protein kinases, culminating in the activation of Fos and hence of the AP-1 transcription factors. The combination of NFκB, NFAT and AP-1 act upon the T-cell chromosomes, initiating new gene transcription that results in the differentiation, proliferation and effector actions of T cells.

> 2   *What is known about the normal functions of the receptors of which the γ_c chain forms a part and how might these account for the phenotype of X-linked SCID?*

From what we know of their functions, defects in the receptors for IL-2, IL-4, IL-9, and IL-15 do not appear relevant to the early block in T-cell development seen in X-linked SCID, because all activate mature lymphocytes or other effector cells. However, the receptor for IL-7 is thought to be important for pre-T-cell growth, and in mice (which are deficient in both B and T cells when they lack the γ_c chain), for pre-B-cell growth as well. Mice with a defect in the IL-7 receptor alone suffer from blocks in T- and B-cell development that resemble those seen in mice lacking the γ_c chain. A loss of IL-7 receptor function is therefore likely to be the most important loss of function responsible for X-linked SCID.

3 *Martin received a haploidentical transplant from his mother. What was the advantage and the disadvantage of using his mother as a donor as opposed to his father?*

In Martin's case, the advantage of choosing his mother as donor lies in the facility of proving the establishment of chimerism. The transplanted T cells, responding to PHA, would have an XX karyotype in this male infant. If the father had been used as a donor this valuable marker would not be available to make it easy to distinguish between the transplanted cells and Martin's own cells. Conversely, half the maternal T-cell precursors will not mature because they carry the genetic defect. This would not be a problem if Martin's father had been used as the donor.

4 *Why was it necessary to treat the maternal marrow donation with a monoclonal antibody to mature T cells and complement prior to the transplant?*

To prevent graft-versus-host disease. The mother's bone marrow donation almost certainly contained mature T cells capable of reacting with the paternal HLA antigens inherited by Martin from his father. Recognition of the paternal HLA antigens in Martin by the T-cell antigen receptors of the maternal T cells would incite graft-versus-host disease. Martin was also treated with a short course of the immunosuppressant drugs cyclosporin and prednisone after receiving the transplant, in order to suppress any graft-versus-host disease that might have arisen.

5 *Frequently infants with SCID get very ill with Pneumocystis carinii pneumonia after a successful transplant. For this reason they are treated prophylactically with antibiotics to get rid of any P. carinii organisms that may be present in their lungs. Why is this a wise therapeutic maneuver and how do you explain the worsening of pneumonia after a transplant?*

The *P. carinii* organisms are present in lung fluid and in the pulmonary macrophages. They do not incite an inflammatory response until the infant has T cells bearing a T-cell antigen receptor for *P. carinii* antigens. After a successful transplant, the infant is rendered chimeric. The transplanted T cells recognize *P. carinii* antigens and incite an inflammatory response, which makes the pneumonia more severe.

6 *We have already discussed the risks associated with giving polio virus vaccine to immunodeficient infants in the case of X-linked agamma-globulinemia. Martin would not have been able to clear the poliovirus he*

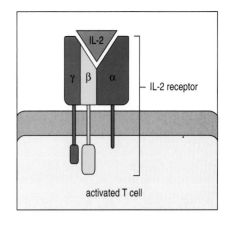

Fig. 5.7 IL-2 receptors are three-chain structures composed of α, β, and γ chains. Resting mature T cells express only the β and γ chains, which bind IL-2 with moderate affinity, allowing resting T cells to respond to very high concentrations of IL-2. Activation of T cells induces the synthesis of the α chain and the formation of the heterotrimeric receptor, which has a high affinity for IL-2 and allows the T cell to respond to very low concentrations of IL-2. The β and γ chains show amino acid similarities to cell-surface receptors for growth hormone and prolactin, all of which regulate cell growth and differentiation. The α chain is also expressed briefly by immature thymocytes as they are rearranging their T-cell receptor β-chain genes. But the normal T cell development seen in mice lacking IL-2 or the IL-2Rβ chain argues against an important function for the complete IL-2R at this developmental stage.

received until he was started on gamma-globulin therapy, but luckily a pathogenic variant did not arise during that time. Fortunately Martin escaped being given any other live vaccines before he was diagnosed. In many countries of the world (but not the United States) infants are universally given bacille Calmette-Guérin (BCG), an attenuated form of the tuberculosis bacillus, which provides partial protection against tuberculosis infection. BCG incites a cell-mediated immune response and after receiving it infants become tuberculin-positive, which means they show a delayed-type hypersensitivity response to a skin-prick with minute quantities of tuberculin. In the United States, the tuberculin test is considered so diagnostically valuable for the detection of new tuberculosis infections that BCG is not given. What do you think happens to infants with SCID after they are given BCG?

They die of progressive BCG infection. This ordinarily harmless attenuated bacillus is a pathogen in individuals with compromised cell-mediated immunity.

7 *Before smallpox was eradicated in the world, vaccinia virus was routinely administered to all children. What happened to infants with SCID who were vaccinated with vaccinia virus?*

They developed a progressive vaccinia infection, which spread contiguously in the skin from the site of inoculation—so-called vaccinia gangrenosa. It proved invariably fatal in these infants. No live vaccines of any kind should be given to an immunodeficient child or adult.

CASE 6 | Adenosine Deaminase Deficiency

The purine degradation pathway and lymphocytes.

In Case 5 we learned about severe combined immunodeficiency (SCID) and examined the most common form—that due to an X-linked mutation in the gamma chain (γ_c) common to several interleukin receptors. In this type of SCID, T cells are virtually absent while B cells are present in normal numbers although they are not functional. Further examination of this phenotype also reveals an absence of natural killer (NK) cells. Thus, X-linked SCID is classified as T⁻B⁺NK⁻ SCID.

SCID patients with an almost complete absence of B cells as well as T cells are also encountered; their phenotype is T⁻B⁻NK⁺. This phenotype is exclusively associated with SCID transmitted as an autosomal recessive condition and has a quite different biochemical cause from X-linked SCID. The most common genetic defect encountered in these patients is mutation in the gene encoding

Topics bearing on this case:

Defects in lymphocyte function result in severe combined immunodeficiency

Bone marrow transplantation

Mixed lymphocyte reaction

Minor histocompatibility antigens

Graft-versus-host reaction

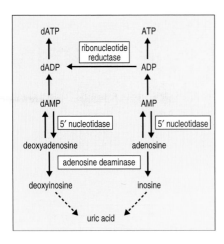

Fig. 6.1 Purine metabolic pathways. Adenosine deaminase catalyzes the conversion of deoxyadenosine and adenosine to deoxyinosine and inosine, which are eventually converted to waste products and excreted. In the absence of ADA in lymphocytes, adenosine and deoxyadenosine metabolites accumulate. In cells other than lymphocytes, the enzyme 5′ nucleotidase can convert AMP and dAMP to adenosine and deoxyadenosine, thus preventing the build-up of potentially toxic metabolites.

the purine degradation enzyme adenosine deaminase (ADA), which leads to no enzymes being produced. ADA is a ubiquitous housekeeping enzyme found in all mammalian cells and in blood serum; it converts the purine nucleosides adenosine and deoxyadenosine to inosine and deoxyinosine respectively, and hence to waste products that are excreted (Fig. 6.1). In its absence, cells can accumulate excessive amounts of adenosine and deoxyadenosine metabolites. As these tend to be toxic in excess one might expect an absence of ADA activity to seriously affect all the cells of the body. Quite surprisingly, however, it turns out that lymphocytes are particularly susceptible to the toxic effects of these metabolites, and thus ADA deficiency leads to SCID.

The case of Roberta Alden: lymphocytes poisoned by toxic metabolites.

3-week-old female infant with thrush. Family history of SCID.

The Aldens are an African-American family from a remote rural area of the state of Georgia. Mr and Mrs Alden are probably distantly related to each other. At the time they moved to Boston they had seven healthy children—four boys and three girls. Their eighth child was a boy, who developed severe pneumonia at 3 months old and died at the City Hospital. An autopsy revealed that he had SCID. His thymus had a fetal appearance with only rare thymocytes and no Hassall's corpuscles (see Fig. 5.2). Two years later the Aldens had a daughter, named Roberta. She appeared healthy at birth but at 3 weeks old she developed deeply pigmented spots on her trunk and face. Three weeks later Mrs Alden noticed thrush in Roberta's mouth (see Fig. 5.4).

Fig. 6.2 Chest radiographs of 5-month-old infants. In the left panel the thymic shadow in a normal healthy child is surrounded by a dashed blue line. In the right panel the heart shadow (surrounded by a dashed red line) is clearly visible in the SCID infant owing to the absence of the thymus. Courtesy of T Griscom.

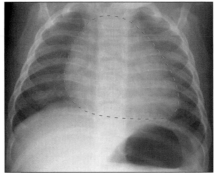
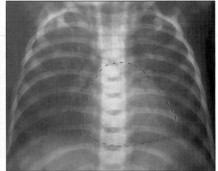

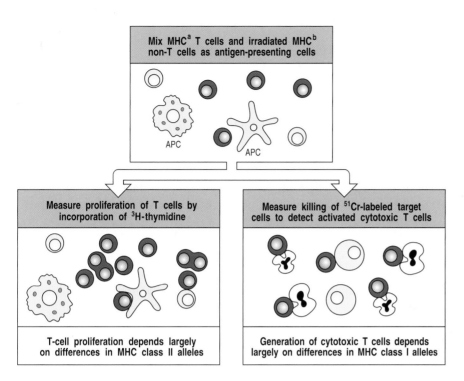

Fig. 6.3 The mixed lymphocyte reaction (MLR) can be used to detect histoincompatibility. Lymphocytes from the two individuals who are to be tested for compatibility are isolated from peripheral blood. The cells from one person (yellow), which also contain antigen-presenting cells, are either irradiated or treated with mitomycin C; they will act as stimulator cells but cannot now respond by DNA synthesis and cell division to antigenic stimulation by the other person's cells. The cells from the two individuals are then mixed (top panel). If the unirradiated lymphocytes (the responders, blue) contain alloreactive T cells, these will be stimulated to proliferate and differentiate to effector cells. Between 3 and 7 days after mixing, the culture is assessed for T-cell proliferation (bottom left panel), which is mainly the result of CD4 T cells' recognizing differences in MHC class II molecules, and for the generation of activated cytotoxic T cells (bottom right panel), which respond to differences in MHC class I molecules. When the MLR is used to select a bone marrow donor, the prospective donor's cells are used as the responder cells and the prospective recipient's cells as the stimulator cells.

Aware of the family history, Roberta's pediatrician ordered a chest x-ray and blood studies. No thymic shadow (Fig. 6.2) was seen in the chest x-ray, and the anterior margins of the ribs were flared. Roberta's lymphocyte count was 150 cells μl^{-1} (normal for an infant is >3000 cells μl^{-1}). Her lymphocytes did not respond to the non-specific T-cell mitogen phytohemagglutinin. A diagnosis of SCID was established.

HLA typing of Roberta, her parents and her seven siblings revealed that her HLA type was identical with one sister, Ellen, and one brother, John. To test directly for histocompatibility, Ellen's and John's blood cells were tested separately against Roberta's in the mixed lymphocyte reaction (Fig. 6.3). No reaction was seen in either case. John and Roberta also had the same blood type (AB) and so John was chosen as the bone marrow donor. Bone marrow cells (2.8×10^7 cells) were removed from John's iliac crest bone and infused into Roberta (this was calculated to be a dose of 5×10^6 cells per kilogram of Roberta's body weight).

Twelve days after the bone marrow transplant, Roberta's lymphocyte count had increased to 500 μl^{-1}, and the response of her blood lymphocytes to phytohemagglutinin had risen to half normal. A karyotype of the responding cells revealed that they had XY sex chromosomes; they were thus of male origin. Two days later Roberta developed a transient graft-versus-host reaction characterized by fever and eosinophilia, which lasted three days. She continued to gain weight and was discharged from the hospital. Several weeks later she, her parents, and all her siblings were affected with a severe influenza-like respiratory infection. She recovered from this without problems and her lymphocyte count rose to 1575 μl^{-1}; she also now had a normal response to phytohemagglutinin. Roberta continued to grow and develop normally and remained free of infections. All her lymphocytes continue to have an XY karyotype.

Diagnosis of SCID. Consider bone marrow transplant.

Bone marrow transplant from HLA- and blood group-identical brother.

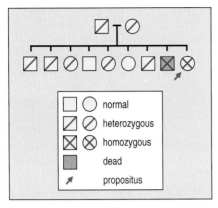

Fig. 6.4 Inheritance of ADA deficiency in Roberta's family. ADA deficiency is inherited as an autosomal recessive condition.

Adenosine deaminase deficiency.

The gene encoding adenosine deaminase (ADA) is located on chromosome 20. SCID due to ADA deficiency appears in homozygotes for defects in the ADA gene that result in an inactive enzyme or no enzyme. As production of ADA from the normal gene in heterozygotes is sufficient to compensate, the disease is inherited as an autosomal recessive condition (Fig. 6.4). Homozygotes for mutations in this gene show the most profoundly lymphopenic form of SCID. The thymus is poorly developed and there is a characteristic abnormality in the rib bones. The pigmented spots that appeared on the trunk and face were the result of a localized graft-versus-host reaction due to maternal cells.

Normally, the adenosine and deoxyadenosine content of cells is limited by ADA, which converts these nucleotides to inosine (and deoxyinosine) and subsequently to urate, which is excreted. The limited amounts of adenosine within cells are converted to adenosine monophosphate (AMP), adenosine diphosphate (ADP), and adenosine triphosphate (ATP). Deoxyadenosine is converted to dAMP, dADP, and dATP (see Fig. 6.1). In the absence of ADA, these metabolites can accumulate in up to 1000-fold excess within cells. Excess dATP in particular inhibits the enzyme ribonucleotide reductase, which is required for the synthesis of all the deoxynucleotides required for DNA synthesis (Fig. 6.5); its inhibition is probably the main culprit in causing the death and non-development of lymphocytes.

The thymus contains 13 times more ADA than any other tissue in the body. As adenosine metabolites are very toxic to lymphocytes, the high level of ADA in the thymus is probably crucial for normal thymocyte development. The reason that lymphocytes are particularly vulnerable to the accumulation of these metabolic poisons is probably because they are relatively deficient in the enzyme 5′ nucleotidase. This enzyme degrades AMP and dAMP to adenosine and deoxyadenosine (see Fig. 6.1) and thereby prevents the excessive accumulation of ADP, ATP, dADP, and dATP, even in the absence of ADA.

The administration of ADA, which is commercially available in a form bound to polyethylene glycol (PEG-ADA), clears the noxious metabolites and results in improved immune function. The same beneficial result has been achieved by gene therapy. The patient's white blood cells are transfected *in vitro* with a retroviral vector bearing a normal ADA gene. They are induced to proliferate *in vitro* by treatment with mitogen and then re-infused into the patient. This treatment is very laborious because it has to be repeated every few months to achieve success.

Another enzyme in the purine degradation pathway, purine nucleoside phosphorylase (PNP), degrades guanosine to inosine. Its absence results in excessive accumulation of the guanosine metabolites GMP, GDP, and GTP, as well as the deoxyguanosine metabolites dGMP, dGDP, and dGTP, which, like the adenosine metabolites, are toxic to lymphocytes. Genetic defects in PNP resulting in SCID have also been found.

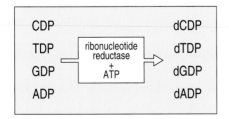

Fig. 6.5 Role of ribonucleotide reductase in generating deoxynucleotides required for DNA synthesis.

Discussion and questions.

1 *The absence of a thymic shadow in the chest radiograph of a young infant can be helpful in the diagnosis of SCID but it is not a reliable finding. Why is this so?*

The thymus gland can shrink (involute) as the result of many kinds of stress, particularly infection.

2 *John and Roberta were histocompatible at the major histocompatibility locus—HLA. Despite this, Roberta sustained a transient graft-versus-host reaction characterized by fever and eosinophilia. How do you explain this?*

The brief, mild graft-versus host reaction was probably due to incompatibility at one or more minor histocompatibility loci.

3 *In a mixed lymphocyte reaction, the cells used to stimulate the response are treated with mitomycin, a mitotic inhibitor, to prevent their responding to the responder cells. When Roberta's and John's cells were tested in a mixed lymphocyte reaction to ascertain whether his T cells would respond to Roberta's cells it was not necessary to treat Roberta's cells with mitomycin. Why was this the case?*

The T cells of a patient with SCID are incapable of responding to a mitogenic stimulus. As Roberta's cells could not respond to John's it was not necessary to add mitomycin to the mixed lymphocyte reaction.

4 *Can you think of a reason for the fact that NK cells are normal in ADA deficiency and are not affected by the metabolic defect?*

NK cells, like most other cells in the body, probably have sufficient 5' nucleotidase to prevent excessive dATP accumulation.

CASE 7 | Toxic Shock Syndrome

Superantigens cause excessive stimulation of T cells and macrophages.

A conventional protein antigen activates only those T lymphocytes that bear a T-cell receptor specific for peptides derived from the antigen, which are a tiny subset of the total T-cell pool. For example, only about 1 in 10,000 circulating T cells from donors immunized with tetanus toxoid subsequently proliferate when stimulated by the toxoid. In contrast, a special class of antigens known as superantigens directly activate large numbers of T cells. Superantigens are bacterial and viral proteins that bind as the whole unprocessed protein to MHC class II molecules outside the peptide-binding groove and also to the lateral face of certain subsets of V_β chains of the T-cell receptor.

Superantigen therefore crosslinks MHC class II molecules on antigen-presenting cells to T-cell receptors with the appropriate V_β specificity (Fig. 7.1). Crosslinking results in the activation of both the T cell and the antigen-presenting cell as a result of signals initiated by the T-cell receptor and the MHC class II molecule respectively. Each superantigen is capable of binding to a limited group of V_β regions. For example, the toxic shock syndrome toxin-1 (TSST-1), which is produced by certain strains of the Gram-positive bacterium *Staphylococcus aureus*, stimulates all those T cells that express the $V_\beta2$ gene segment. As there is a limited repertoire of V_β gene segments, any superantigen will stimulate between 2% and 20% of all T cells.

This mode of stimulation is not specific for the pathogen and thus does not lead to adaptive immunity. Instead, it causes excessive production of cytokines by the large number of activated CD4 T cells, the predominant responding population. This massive cytokine release has two effects on the host: systemic toxicity, which manifests itself as toxic shock syndrome, and suppression of the adaptive immune response. Both these effects contribute to the pathogenicity of microbes that produce superantigens.

Topics bearing on this case:
T-cell activation
Superantigens
Macrophage activation
Toxic shock
Cytokines

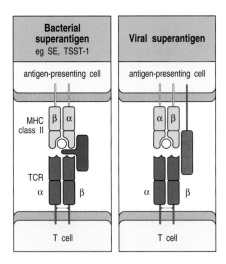

Bacterial superantigen eg SE, TSST-1	Viral superantigen

Fig. 7.1 Superantigens bind directly to T-cell receptors and MHC molecules. Superantigens interact with MHC class II molecules and T-cell receptors in a way that is quite distinct from the way that normal peptide antigens bind. Superantigens bind independently to MHC class II molecules and to T-cell receptors, binding to the V_β domain of the T-cell receptor, away from the peptide-binding site. The T-cell receptor α chain is not directly involved in binding superantigen.

The case of Claire Bourbon: life-threatening shock from a superantigen.

Claire Bourbon was a healthy 16-year-old with a history of mild asthma and allergic rhinitis who suddenly became ill with a fever, general muscle aches and dizziness. She felt nauseous and vomited. Claire's temperature rose to 39.8°C and her mother rushed her to the family physician. En route she briefly lost consciousness, and a red rash developed over her arms and spread rapidly to most of the body.

Upon arrival at the physician's she appeared quite ill and was immediately referred to the Emergency Department. She was alert but listless and her general condition gave cause for concern. On examination, her temperature was 37.8°C, and heart rate and respiration rate were markedly elevated, at 140 beats per minute and 24 breaths per minute respectively. Blood pressure was depressed—98/67 lying supine, 83/49 when seated, and 67/25 when standing—and showed evidence of significant volume depletion.

A bright red rash of flat and raised lesions was apparent on her trunk and extremities, but there were no petechiae (small subcutaneous hemorrhages) and no signs of localized infection.

16-year-old female; systemic shock and bright red rash.

Questioning revealed that Claire had not taken alcohol or drugs and had not been exposed to other ill individuals. Her last menstrual period had been 6 weeks before, and she had developed vaginal bleeding on the day previous to the onset of her illness. She had not used a tampon overnight, but had inserted one that morning, before she became ill.

Given her critical status, extensive laboratory tests were carried out. Her white blood cell count was raised, at 21,000 cells μl^{-1}, with a predominance of neutrophils and band forms (immature neutrophils), indicating increased mobilization of neutrophils from the bone marrow. Serum electrolyte levels were within normal limits. The blood coagulation time was slightly prolonged and serum transaminase levels were raised; both of these signs are consistent with abnormal liver function.

Cerebrospinal fluid (CSF) was normal and showed no evidence of infection. Cultures of blood, urine, CSF, and vaginal fluid were made and Claire was given a cephalosporin antibiotic (ceftriaxone) along with 2 liters of fluid intravenously. Her blood pressure improved and she was immediately admitted to the intensive care unit, where she developed petechiae. She was treated with intravenous fluids, two anti-staphylococcal antibiotics (oxacillin and clindamycin) as well as a cephalosporin (cefotaxime) with gradual improvement in her overall condition.

started period day before? toxic shock syndrome

Her blood, urine, and CSF cultures remained sterile, but her vaginal culture was positive for abundant *S. aureus*. She was subsequently transferred to the regular in-patient ward and treated for 7 days with anti-staphylococcal antibiotics. The rash gradually faded.

Toxic shock syndrome.

Claire suffered from staphylococcal toxic shock syndrome (TSS), a striking example of the dramatic physiologic alterations caused by superantigens. TSS is a serious disease characterized by rapid onset of fever, a rash, organ failure, and shock. The majority of cases occur in menstruating women, typically in their teenage years, but cases do occur in all age groups. TSS is typically associated with a localized *S. aureus* infection (for example, subcutaneous abscesses, osteomyelitis, and infected wounds), staphylococcal food poisoning, or local colonization, as occurred in the vagina in this case. When kept in the vagina for a long time (>12 hours) tampons soaked with menstrual fluids can enhance the growth of the bacteria which are the source of superantigens. It is unlikely that the tampon played a part in Claire's disease as it was inserted less than 6 hours before the onset of symptoms. As well as TSST-1, toxigenic strains of *S. aureus* can produce other enterotoxins (such as enterotoxin B) that act as superantigens, with similar clinical consequences. In addition, microorganisms other than *S. aureus* secrete superantigens that can cause disease (Fig. 7.2).

Fig. 7.2 Superantigens, V_β usage and disease.

Disease	Superantigen	TCR V_β
Definite role for superantigen		
Toxic shock syndrome	TSST-1	$V_\beta2$
Staphylococcal food poisoning	SEA	$V_\beta3$, $V_\beta11$
	SEB	$V_\beta3$, $V_\beta12$, $V_\beta14$, $V_\beta15$, $V_\beta17$, $V_\beta20$
	SEC	$V_\beta5$, $V_\beta12$, $V_\beta13.1–2$, $V_\beta14$, $V_\beta15$, $V_\beta17$, $V_\beta20$
	SED	$V_\beta5$, $V_\beta12$
	SEE	$V_\beta5.1$, $V_\beta6.1–3$, $V_\beta8$, $V_\beta18$
Streptococcal toxic shock syndrome	SPE-A	$V_\beta8$, $V_\beta12$, $V_\beta14$
Scarlet fever	SPE-B	$V_\beta2$, $V_\beta8$
Mycoplasma arthritidis (rodent)	MAM	$V_\beta17$
Clostridium perfringens	Enterotoxin	$V_\beta6.9$, $V_\beta22$
Suspected role for superantigen		
HIV	CMV	$V_\beta12$
Type I diabetes mellitus	MMTV-like	$V_\beta7$
Rabies virus	Nucleocapsid	$V_\beta8$
Toxoplasmosis	?	$V_\beta5$
Mycobacterium tuberculosis	?	$V_\beta8$
Yersinia enterocolitica	?	$V_\beta3$, $V_\beta6$, $V_\beta11$
Kawasaki disease	?	$V_\beta2$, $V_\beta8$

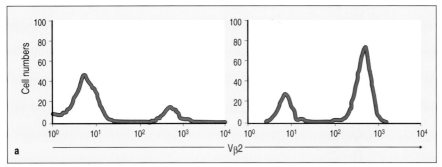
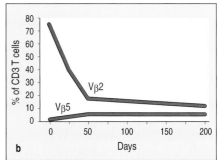

Fig. 7.3 Expansion in numbers of V$_\beta$2 T cells in toxic shock syndrome (TSS). Panel a, FACS analysis. Peripheral blood T cells from a normal control (left) and a patient with acute TSS (right). Cells are stained with anti-V$_\beta$2 monoclonal antibody with a fluorescein tag. There is an increased percentage of V$_\beta$2 T cells in the patient. The horizontal axis represents the mean fluorescence intensity. Panel b, time course of persistence of high numbers and return to normal of V$_\beta$2 T cells in a patient with TSS.

Consistent with the V$_\beta$2 specificity of TSST-1, examination of the circulating lymphocytes from patients in the acute phase of TSS typically reveals a much higher proportion of circulating V$_\beta$2 T cells compared with cells using other V$_\beta$ segments. As the illness resolves there is a gradual return to near normal proportions. The expansion in the numbers of V$_\beta$2 cells can be measured by examining the surface expression of T-cell receptors containing a V$_\beta$2 region using immunofluorescence (Fig. 7.3) or by semiquantitative measurement of mRNA transcripts encoding for V$_\beta$2 T-cell receptor chains using reverse transcription and the polymerase chain reaction (RT-PCR) (Fig. 7.4).

Although all the T cells activated by a given superantigen share a common V$_\beta$ region, they will differ in their specificity for conventional peptide antigens. Sequencing of the T-cell receptor from superantigen-activated T cells reveals a different use of D and J gene segments by the β chains and a wide diversity of α chains. These receptors will encompass a wide variety of antigen specificities. In contrast to superantigen, conventional antigen induces the clonotypic expansion of T cells. All the T cells in any given clone will have identical D and J gene segments in their β chains and identical α chains. Because the pool of V$_\beta$-restricted T cells activated by superantigen may contain autoreactive T cells, it has been postulated that superantigens may trigger autoimmune disease.

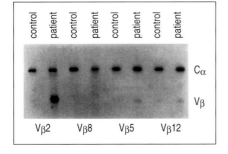

Fig. 7.4 RT-PCR analysis of T-cell receptor V$_\beta$ mRNA. Autoradiograms of T-cell receptor chain transcripts amplified by reverse transcription followed by polymerase chain reaction (RT-PCR). T cells from a patient with toxic shock syndrome and a control individual were stimulated with anti-CD3 antibody and IL-2 before the extraction of RNA and generation of cDNA. Each reaction contained specific oligonucleotide primers to expand the particular V$_\beta$ gene segment indicated (170–220 base pairs), as well as a C$_\alpha$ gene segment (approx. 600 bp) as a control to ascertain that equivalent amounts of mRNA were used. Photograph courtesy of Y Choy.

Many of the manifestations of TSS are the result of massive and unregulated cytokine production triggered by the activation of immune system cells. TSST-1 is more effective than bacterial lipopolysaccharide in inducing the synthesis and secretion of interleukin (IL)-1 and tumor necrosis factor (TNF)-α by monocytes. TSST-1 is also a potent T-cell mitogen for those T cells whose receptors it engages; it also induces them to produce large amounts of cytokines, including IL-2 and interferon (IFN)-γ.

IL-1 and TNF-α are critical in the induction of the acute-phase response and induce fever and the production of IL-6. IL-1 and TNF-α also activate vascular endothelium and, together with IL-2, increase vascular permeability with the subsequent leakage of fluid from the intravascular space into the perivasculature. It is these effects of the massive overproduction of TNF-α that result in the toxic shock: edema and intravascular volume depletion leading to hypotension and shock with multiple organ failure.

Susceptibility to TSS seems to correlate with a poor antibody response to TSST-1. While the majority of healthy individuals have protective antibody titers to TSST-1, more than 80% of patients with TSS lack anti-TSST-1 antibodies during the acute illness, and most fail to develop anti-TSST-1 antibodies following convalescence. Possible explanations include an inability on the patient's part to mount an antibody response against TSST-1 and staphylococcal enterotoxins, or a specific inhibition of such a response by the toxins.

Discussion and questions.

1   *How do you determine whether a protein behaves as a superantigen?*

Superantigens, but not conventional antigens, can activate naive T cells. Superantigens will thus induce the proliferation of lymphocytes from newborns and from the thymus, as previous exposure to the antigen and expansion of the number of antigen-reactive cells is not required. Superantigens do not require processing by accessory cells and are thus able to induce the proliferation of purified T cells in the presence of paraformaldehyde-treated monocytes, which lack the capacity to process antigen. Direct binding of a labeled protein to cells positive for MHC class II, or its co-precipitation with MHC class II molecules, confirms it as a superantigen.

2   *Explain the rapid progression of clinical symptoms following introduction of superantigen compared with the delay in apparent responses to conventional antigen.*

During the evolution of an adaptive immune response to conventional antigen, a cascade of events must occur over a relatively long period of time. The antigen has to be internalized, processed, and presented as peptide:MHC complexes by antigen-presenting cells. The complexes are recognized only by those T cells bearing a T-cell receptor specific for the antigen-derived peptides— a fraction of a percent ($<0.1\%$) of the entire T-cell pool. These few antigen-specific T cells must then proliferate and bystander cells be recruited before an effective response can be mounted. In contrast, superantigen-induced immune activation is independent of antigen processing, thus bypassing the first step, and immediately activates a sizeable fraction of T cells. A very small number of superantigen molecules is sufficient to activate a T cell with the appropriate V_β region in its receptor (<10 molecules per T cell). Activation results in a massive secretion of T-cell cytokines, which include IL-2, IFN-γ, TNF-α, and TNF-β. In addition, superantigens can directly activate monocytes and dendritic cells by crosslinking their surface MHC class II molecules. Crosslinking is effected by superantigens bound to T-cell receptor β chains and/or because a number of superantigens, including TSST-1, have two distinct binding sites for MHC class II molecules. Crosslinking of MHC class II molecules causes a rapid and massive release of cytokines such as IL-1, TNF-α, IL-6, IL-8, and IL-12. This is associated with the upregulation of B-7 co-stimulatory molecules on these cells, which, together with cytokine action, further amplifies T-cell activation by superantigen. Thus, minute amounts of superantigen are sufficient to rapidly activate a large number of T cells and monocytes/macrophages, resulting in an amplificatory loop and in a massive outpouring of cytokines, which leads to the rapid appearance of clinical symptoms.

3 | *What are the potential mechanisms of liver injury in TSS?*

Liver injury may occur as a result of decreased organ perfusion during hypotension. However, immunologic mechanisms may also contribute to injury. Hepatocytes express Fas, a cell-surface molecule crucial for the induction of apoptosis (programmed cell death). T-cell activation by superantigens and the massive release of cytokines results in the upregulation of the natural ligand for Fas—FasL—on the surface of circulating lymphocytes. Crosslinking of Fas on hepatocytes by FasL on circulating lymphocytes results in the triggering of apoptosis in hepatocytes. In addition, circulating cytokines such as TNF-α are also capable of triggering cell death and can result in liver injury.

4 | *Is Claire susceptible to another bout of TSS?*

Protection against toxic shock is conferred by antibodies against the super-antigen, which neutralize it before it can cause disease. In order to stimulate an antibody response, the superantigen must be recognized, internalized, and processed by superantigen-specific B cells which then present the antigenic peptides to antigen-specific T cells. These are activated to become helper T cells that can in turn stimulate the production of superantigen-specific antibodies on re-exposure to the superantigen. Antibodies to other antigens that cross-react with the superantigen may also confer protection.

In humans, there is evidence that during and following TSST-associated illness, $V_\beta 2$ T cells become anergic and thus cannot provide help to superantigen-specific B cells. Patients with TSS therefore fail to develop TSST-1-specific antibody. So Claire is, unfortunately, likely to be at risk of another episode of TSS. Hopefully, she will eventually develop anti-TSST-1 antibodies.

CASE 8 | Contact Sensitivity to Poison Ivy

A delayed hypersensitivity reaction to a hapten.

Allergic or hypersensitivity reactions can be elicited by antigens that are not associated with infectious agents, for example pollen, dust, food, and chemicals in the environment. They do not usually occur on the first encounter with the antigen, but a second or subsequent exposure of a sensitized individual causes an allergic reaction whose symptoms will depend on the type of antigen, the route by which it enters the body, and the cells involved in the immune response. These unwanted responses can cause distressing symptoms, tissue damage, and even death. These are the same reactions that would be provoked by a pathogenic antigen that behaved in the same way. When they are not helping to clear up an infection, however, these damaging side-effects are clearly unwanted.

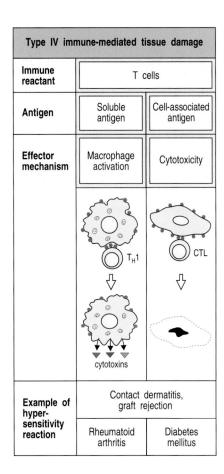

Type IV immune-mediated tissue damage		
Immune reactant	T cells	
Antigen	Soluble antigen	Cell-associated antigen
Effector mechanism	Macrophage activation	Cytotoxicity
	T_H1	CTL / cytotoxins
Example of hyper-sensitivity reaction	Contact dermatitis, graft rejection	
	Rheumatoid arthritis	Diabetes mellitus

Fig. 8.1 There are four types of immune-mediated tissue damage. Types I–III are antibody-mediated and are distinguished by the different types of antigens recognized and the different classes of antibody involved. We shall see examples of type I responses in Case 16 and Case 24 and of type III in Case 25. Type IV reactions are T-cell mediated and can be subdivided into two classes. In the first class, tissue damage is caused by T_H1 cells, which activate macrophages, leading to an inflammatory response. On encounter with antigen, effector T_H1 cells secrete cytokines, such as interferon-γ, that activate macrophages, and to a lesser extent mast cells, to release cytokines and inflammatory mediators that cause the symptoms. In the second class of type IV reactions, damage is caused directly by cytotoxic T cells that attack tissue cells presenting the sensitizing antigen on their surface. The delay in the appearance of a type IV hypersensitivity reaction is due to the time it takes to recruit antigen-specific T cells and other cells to the site of antigen localization and to develop the inflammatory response. This is in contrast to the IgE-mediated type I allergic reactions, where the mast-cell degranulation that causes many of the symptoms is achieved by native antigen directly cross-linking the pre-formed IgE molecules on the surface of the mast cells. These reactions occur within 20–30 minutes of contact with antigen. Because a delayed hypersensitivity response involves antigen processing and presentation to achieve T-cell activation, quite large amounts of antigen need to be present at the site of contact. The amount of antigen required is two or three orders of magnitude greater than that required to initiate an immediate allergic reaction.

There are four main types of hypersensitivity reactions, which are distinguished by the type of immune cells and antibodies involved, and the pathologies produced. The one discussed here is an example of a type IV (delayed hypersensitivity) reaction (Fig. 8.1). Many allergic reactions occur within minutes or a few hours of encounter with the antigen, but some take a day or two to appear (Fig. 8.2). These are the delayed hypersensitivity reactions (type IV hypersensitivity). Delayed hypersensitivity reactions are mediated by T cells only, either T_H1 CD4 T cells or cytotoxic CD8 T cells, or sometimes both. Antibodies are not involved. The reactions can be triggered by foreign proteins or by self proteins that have become modified by attachment of a hapten, such as a small organic molecule or metal ions. A common type of delayed hypersensitivity reaction is contact dermatitis, a skin rash caused by direct contact with the antigen.

Fig. 8.2 The time course of a delayed-type hypersensitivity reaction. The first phase involves uptake, processing, and presentation of the antigen by local antigen-presenting cells. In the second phase, T_H1 cells that were primed by a previous exposure to the antigen migrate into the site of injection and become activated. Because these specific cells are rare, and because there is no inflammation to attract cells into the site, it may take several hours for a T cell of the correct specificity to arrive. These cells release mediators that activate local endothelial cells, recruiting an inflammatory cell infiltrate dominated by macrophages and causing the accumulation of fluid and protein. At this point, the lesion becomes apparent.

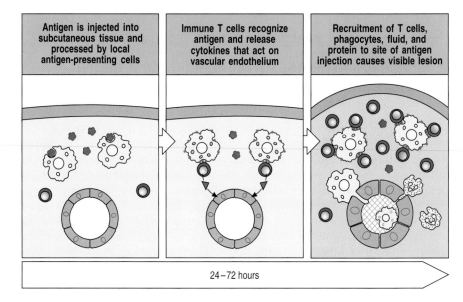

Antigen is injected into subcutaneous tissue and processed by local antigen-presenting cells	Immune T cells recognize antigen and release cytokines that act on vascular endothelium	Recruitment of T cells, phagocytes, fluid, and protein to site of antigen injection causes visible lesion

24–72 hours

Delayed hypersensitivity reactions fall into two classes (see Fig. 8.1). In the first, the damage is due to an inflammatory response and tissue destruction by T_H1 cells and the macrophages they activate. In the second class of delayed hypersensitivity reactions, tissue damage is caused mainly by the direct action of antigen-specific cytotoxic CD8 T cells on target cells displaying the foreign antigen. Some antigens may cause a combination of both types of reaction.

This case describes the most frequently encountered delayed hypersensitivity reaction in the United States—contact dermatitis due to the woodland plant poison ivy.

The case of Paul Stein: a sudden appearance of a severe rash.

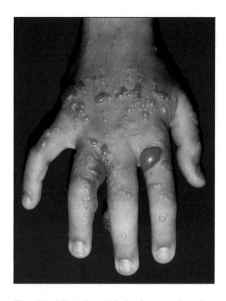

Fig. 8.3 Blistering skin lesions on hand of patient with poison ivy contact dermatitis.

Paul Stein was 7 years old and had enjoyed perfect health until 2 days after he returned from a hike with his summer camp group, when itchy red skin eruptions appeared all along his right arm. Within a day or two, the rash had spread to his trunk, face, and genitals. His mother gave him the antihistamine Benadryl (diphenhydramine hydrochloride) orally to suppress the itching, but this only gave partial relief. The rash did not improve and a week after it first appeared he attended the dermatology clinic at the Children's Hospital.

Physical examination revealed large patches of raised, red, elongated, blisters, oozing a little clear fluid, on his body and extremities (Fig. 8.3). Paul also had swollen eyelids and a swollen penis. There was no history of fever, fatigue, or any other symptom. A contact sensitivity reaction to poison ivy was diagnosed.

He was given a corticosteroid-containing cream to apply to the skin lesions three times a day, and Benadryl (25 mg) to take orally three times a day. He was asked to shampoo his hair, wash his body thoroughly with soap and water, and cut his nails short.

Two days later, his parents reported that, although no new eruptions had appeared, the old lesions were not significantly better. Paul was then given the corticosteroid prednisone orally, starting at a dose of 2 mg kg^{-1} body weight per day, which was gradually decreased over a period of 2 weeks. The topical steroid cream was discontinued.

Within a week, the rash had almost disappeared. Upon stopping the prednisone there was a mild flare-up of some lesions and this was controlled by application of topical steroid for a few days. Paul was shown how to identify poison ivy in order to avoid further contact with it, and told to wear long pants and shirts with long sleeves on any future hikes in the woods.

Contact sensitivity to poison ivy.

The reaction to poison ivy is the most commonly seen delayed hypersensitivity reaction in those parts of the United States where the plant grows wild. The absence of fever or general malaise accompanying the rash, and Paul's otherwise excellent health except for the skin lesions, point to a contact sensitivity reaction rather than to a viral or bacterial infection or some other underlying

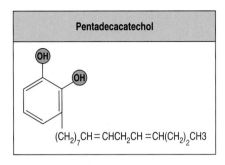

Pentadecacatechol

$(CH_2)_7CH = CHCH_2CH = CH(CH_2)_2CH3$

Fig. 8.4 The chemical formula of pentadecacatechol, the causative agent of contact sensitivity to poison ivy.

long-term illness. The appearance of the rash just 2 days after Paul returned from a hiking trip where he could easily have been in contact with poison ivy virtually clinches the diagnosis.

Contact dermatitis due to poison ivy is caused by a T-cell response to a chemical in the leaf called pentadecacatechol (Fig. 8.4). On contact with the skin, this small, highly reactive, lipid-like molecule penetrates the outer layers, and binds covalently and non-specifically to proteins on the surfaces of skin cells, in which form it functions as a hapten. Most people are susceptible and sensitivity, once acquired, is life-long.

On the very first contact with the plant, the individual becomes sensitized. The haptenated self proteins are ingested by specialized phagocytic cells in the skin (Langerhans' cells) into intracellular vesicles, where they are cleaved into peptides. Some of these peptides will have hapten attached. The peptides bind to MHC class II molecules in the vesicles and are presented as MHC:peptide complexes on the Langerhans' cell surface. If these Langerhans' cells reach a lymph node, they become antigen-presenting cells that can activate naive hapten-specific T cells to become recirculating hapten-specific effector T_H1 cells. The activated effector T_H1 cells can then react with haptenated peptides presented by Langerhans' cells and skin macrophages at the site of contact with the plant to initiate a local delayed hypersensitivity reaction in the skin. As the haptenated peptides can persist in the skin for days, it is possible to get a hypersensitivity reaction to poison ivy even after touching it just once.

The appearance of Paul's rash 2 days after his suspected exposure to poison ivy is typical of a delayed hypersensitivity reaction. The haptenated self peptides presented on skin macrophages and Langerhans' cells at the site of contact with poison ivy are initially recognized by the small number of activated hapten-specific T_H1 cells within the pool of recirculating T cells. On encounter with the haptenated peptides, these T_H1 cells release chemokines (cell-attractant molecules), cytokines, and cytotoxins that initiate an inflammatory reaction and also kill cells directly (Fig. 8.5).

One of the cytokines produced by the T_H1 cells is interferon-γ (IFN-γ), whose main effect in this context is to activate macrophages. The subsequent macrophage activity causes many of the symptoms of the delayed hypersensitivity reaction. Macrophages activated by IFN-γ release cytokines and inflammatory mediators such as interleukins, prostaglandins, nitric oxide (NO), and leukotrienes. The combined effects of T-cell and macrophage activity cause a local inflammatory response and tissue damage at the site of the contact with poison ivy.

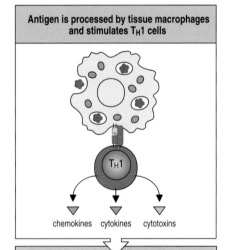

Antigen is processed by tissue macrophages and stimulates T_H1 cells

T_H1

chemokines   cytokines   cytotoxins

IFN-γ	TNF-α/TNF-β
Activates macrophages increasing release of inflammatory mediators	Local tissue destruction. Increased expression of adhesion molecules on local blood vessels

Chemokines	IL-3/GM-CSF
Macrophage recruitment to site of antigen	Monocyte production by bone marrow stem cells

Fig. 8.5 The delayed-type (type IV) hypersensitivity response is directed by cytokines released by T_H1 cells stimulated by antigen. Antigen in the local tissues is processed by antigen-presenting cells and presented on MHC class II molecules. Antigen-specific T_H1 cells can recognize the antigen locally at the site of injection, and release chemokines and cytokines that recruit macrophages to the site of antigen deposition. Antigen presentation by the newly recruited macrophages then amplifies the response. T cells may also affect local blood vessels through the release of TNF-α and TNF-β and stimulate the production of macrophages through the release of IL-3 and GM-CSF (granulocyte–macrophage colony-stimulating factor). Finally, T_H1 cells activate macrophages through the release of IFN-γ, and kill macrophages and other sensitive cells through the release of TNF-β or by the expression of Fas ligand.

The red, raised, blistering skin lesions of poison ivy dermatitis are due to the infiltration of large numbers of blood cells into the tissue at the site of contact, combined with the localized death of tissue cells and the destruction of the extracellular matrix that holds the layers of skin together. One of the first actions of effector T_H1 cells on contact with their antigen is to release the cytokines TNF-α and TNF-β (lymphotoxin), and chemokines such as RANTES. TNF-α in particular increases the expression of adhesion molecules on the endothelium lining near post-capillary venules and increases vascular permeability so that macrophages and other leukocytes adhere to the sides of the blood vessel. This aids their migration from the bloodstream into the tissues in response to the secreted chemokines.

Once at the contact site, macrophages are activated and themselves release cytokines and other inflammatory mediators, which attract more monocytes, T cells, and other leukocytes to the site, thus helping to amplify and maintain the inflammatory reaction. The blood vessels also dilate, which causes the redness associated with the rash. The inflammatory mediators also act on mast cells to cause degranulation and the release of histamine, which is the main cause of the itching that accompanies the reaction.

Tissue destruction, which is a feature of delayed hypersensitivity reactions, is caused both by cytokines and by direct cell–cell interactions. The TNF-α and TNF-β released by T_H1 cells and macrophages act at the same TNF receptors, which are expressed on virtually all types of cell, including skin cells. Stimulation of these receptors induces a 'suicide' pathway in the cells—apoptosis—which causes their death. Activated CD4 T cells also express the Fas ligand (FasL) in their plasma membrane, which interacts with the ubiquitously expressed cell-surface molecule, Fas, to cause the death of the target cell by apoptosis. Other mediators released by activated T cells, such as the enzyme stromelysin, degrade the proteins of the extracellular matrix, that maintain the integrity of the skin.

Lipid-like haptens such as pentadecacatechol can also cause the priming and activation of cytotoxic CD8 T cells, as small fat-soluble molecules can enter the cytosol of skin cells directly by diffusing through the plasma membrane. Once inside, pentadecacatechol binds to intracellular proteins. Peptides generated from the haptenated proteins in the cytosol are delivered to the cell surface associated with MHC class I molecules. These target cells are recognized and attacked by antigen-specific CD8 cytotoxic T cells, which have become primed and activated on a previous encounter with the antigen. The outcome of all these reactions are the raised, red, weeping blisters characteristic of sensitivity to poison ivy.

Corticosteroids are the standard treatment for hapten-mediated contact sensitivity as they inhibit the inflammatory response by inhibiting the production of many of the cytokines and chemokines. Corticosteroids are lipid-like molecules that can diffuse freely across plasma membranes. Once inside the cell, they bind to receptor proteins in the cytoplasm. The receptor:steroid complex enters the nucleus, where it controls the expression of a number of genes. Of relevance here, it induces the production of an inhibitor of the transcription factors required to switch on transcription of the cytokine and chemokine genes. In mild cases of poison ivy dermatitis, topical steroids applied locally are sufficient. In more severe cases such as Paul's, oral steroids are needed to achieve a concentration necessary to inhibit the inflammatory response.

Paul was given antihistamines to block the histamine receptors and counteract the action of the histamine released from mast cells. Antihistamines also counteract itching caused by substances other than histamine (eg prostaglandins released from macrophages).

Discussion and questions.

1 *Paul had lesions not only on the exposed skin but also on areas that would be covered, like the trunk and penis and in areas that were not in obvious contact with poison ivy leaves. How do you explain that?*

Pentadecacatechol can be transferred from the initial point of contact to other areas of the skin by the fingernails after scratching the itchy lesion at the primary site of hapten introduction. This is why it is essential to cut the fingernails short and thoroughly wash off the skin and scalp to remove the chemical and prevent further spread.

2 *How do you explain the recurrence of the lesions after discontinuation of the corticosteroids?*

The half-life of some of the proteins haptenated by pentadecacatechol can be quite long. CD4 memory T cells will continue to be activated as long as the haptenated peptides are being generated. In Paul's case this went beyond the third week following contact with poison ivy.

3 *Paul must take great care to avoid poison ivy all his life, as subsequent reactions to it could be even more severe. Why would this be?*

Once an individual is sensitized, the reaction often becomes worse with each exposure, as each re-exposure not only produces the hypersensitivity reaction but generates more effector and memory T cells. Memory T cells that mediate delayed hypersensitivity reactions, such as contact sensitivity to poison ivy and the tuberculin test, can persist for most of the life of the individual.

4 *How would you confirm that Paul's contact dermatitis was caused by poison ivy rather than by another chemical such as the one found in the leaves of poison sumac (Toxicodendron vernix), another plant that gives rise to contact dermatitis?*

You could perform a patch test. In this test a patch of material impregnated with the hapten is applied to the skin under seal for 48 hours. The area is then examined for redness, swelling, and vesicle formation. Alternatively, peripheral blood mononuclear cells can be incubated with the hapten and T-cell proliferation assessed 6–9 days later.

5 *One of Paul's friends, Brian, has X-linked agammaglobulinemia. What is the likelihood that this boy will develop poison ivy sensitivity?*

The risk of Brian's developing poison ivy sensitivity is at least as high as that for a normal child. This is because antibody plays no discernible role in the genesis of delayed hypersensitivity reactions and T-cell function is normal in X-linked agammaglobulinemia. In fact, clinical observations suggest that boys with X-linked agammaglobulinemia may develop more severe forms of poison ivy. It has been suggested that, in the absence of antibody, more hapten is available for conjugation with self proteins and that, in the absence of antigen presentation by B cells, the T-cell response is skewed more towards T_H1 cells.

6 Delayed hypersensitivity reactions are a rapid, inexpensive, and easy measure of T-cell function. What antigens might you use to test people for T-cell function in this way?

The artifically induced tuberculin reaction is a good model of a delayed hypersensitivity reaction. This skin test detects infection with the bacterium *Mycobacterium tuberculosis*, or previous immunization against tuberculosis with the attenuated live vaccine BCG. Small amounts of tuberculin, a protein derived from *M. tuberculosis*, are injected subcutaneously; a day or two later, a sensitized person develops a small, red, raised area of skin at the site of injection. In countries where BCG is administered routinely to babies, the tuberculin test can be used to test for T-cell function. This is because antigen-specific memory T cells are long-lived, and the sensitivity to tuberculin will persist throughout life. In the USA, children are not immunized with BCG. However, they all receive a full course of diphtheria and tetanus vaccines, which in each case contain purified protein toxoids as the antigen. Contact sensitivity to these two antigens can be used to test T-cell function. Alternatively, antigen derived from the yeast-like fungus *Candida albicans*, which is a normal inhabitant of the body flora, can be used to induce a delayed hypersensitivity reaction in the skin.

7 What are some common causes of contact sensitivity?

Some of the commoner environmental causes of delayed hypersensitivity reactions are insect bites or stings, which introduce insect venom proteins under the skin, and skin contact with chemicals in the leaves of some plants, or with metals such as nickel, beryllium, and chromium (Fig. 8.6). Nickel sensitivity is quite common and often occurs at the site of contact with nickel-containing jewelry. Contact sensitivity to beryllium has been well documented in factory workers engaged in manufacturing fluorescent light bulbs. Celiac disease is a type of delayed sensitivity reaction seen in people who are allergic to the protein gliadin, a constituent of wheat grains and flour. Patients with celiac disease thus have to avoid all food products containing wheat flour.

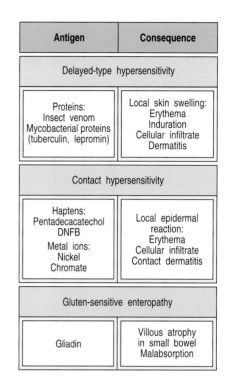

Antigen	Consequence
Delayed-type hypersensitivity	
Proteins: Insect venom Mycobacterial proteins (tuberculin, lepromin)	Local skin swelling: Erythema Induration Cellular infiltrate Dermatitis
Contact hypersensitivity	
Haptens: Pentadecacatechol DNFB Metal ions: Nickel Chromate	Local epidermal reaction: Erythema Cellular infiltrate Contact dermatitis
Gluten-sensitive enteropathy	
Gliadin	Villous atrophy in small bowel Malabsorption

Fig. 8.6 Some type IV hypersensitivity reactions. Depending on the source of antigen and its route of introduction, these clinical conditions have different names and consequences.

CASE 9 | Hereditary Angioneurotic Edema

Regulation of complement activation.

Complement is a system of plasma proteins that participates in a cascade of reactions, generating active components that allow pathogens and immune complexes to be destroyed and eliminated from the body. Complement activation is generally confined to the surface of pathogens or circulating complexes of antibody bound to antigen.

Complement is normally activated by one of three routes—the classical pathway, which is triggered by antigen:antibody complexes or antibody bound to the surface of a pathogen; the lectin-dependent pathway, which is activated by cytokines released by macrophages; and the alternative pathway, in which complement is activated spontaneously on the surface of some bacteria. The early part of each pathway is a series of proteolytic cleavage events leading to the generation of a convertase, a serine protease that cleaves complement component C3 and thereby initiates the effector actions of complement. The C3 convertases generated by the three pathways are different, but evolutionarily homologous, enzymes. Complement components and activation pathways, and the main effector actions of complement, are summarized in Fig. 9.1.

The principal effector molecule, and a focal point of activation for the system, is the large cleavage fragment of C3,C3b. If active C3b, or the homologous but less potent C4b, accidentally becomes bound to a host cell surface instead of a pathogen, the cell can be destroyed. This is usually prevented by the rapid hydrolysis of active C3b and C4b if they do not bind immediately to the surface where they were generated. Protection against inappropriate activation of complement is also provided by regulatory proteins.

One of these, and the most potent inhibitor of the classical pathway, is the C1 inhibitor (C1INH). This belongs to a family of serine protease inhibitors (called serpins) that together constitute 20% of all plasma proteins. In addition to being the sole known inhibitor of C1, C1INH contributes to the regulation of serine proteases of the clotting system and of the kinin system, which is activated by injury to blood vessels and by some bacterial toxins.

Topics bearing on this case:
Classical pathway of complement activation
Inhibition of C1 activation
Alternative pathway of complement activation
Inflammatory effects of complement activation
Regulation of C4b

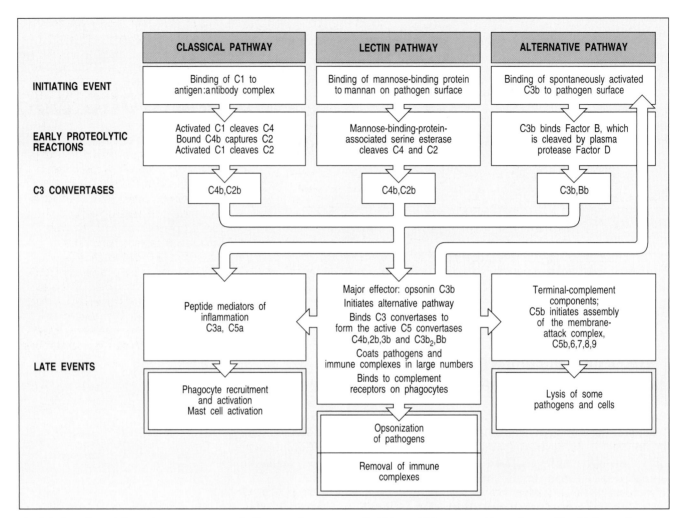

Fig. 9.1 Overview of the main components and effector actions of complement. The early events of all three pathways of complement activation involve a series of cleavage reactions culminating in the formation of an enzymatic activity called a C3 convertase, which cleaves complement component C3. This is the point at which the three pathways converge and the effector functions of complement are generated. The larger cleavage fragment of C3 (C3b) binds to the membrane and opsonizes bacteria, allowing phagocytes to internalize them. The small fragments of C5 and C3, called C5a and C3a, are peptide mediators of local inflammation; vascular permeability is increased and mast cells are recruited and activated to enhance the clearance of pathogens opsonized with C3b. Finally, the C3b bound to the C3 convertase binds C5, allowing the C3 convertase to generate C5b, which associates with the bacterial membrane and triggers the late events, in which the terminal components of complement assemble into a membrane-attack complex that can damage the membrane of certain pathogens.

C1INH intervenes in the first step of the complement pathway, when C1 binds to immunoglobulin molecules on the surface of a pathogen or antigen:antibody complex (Fig. 9.2). Binding of two or more of the six tulip-like heads of the C1q component of C1 is required to trigger the sequential activation of the two associated serine proteases, C1r and C1s. C1INH inhibits both of these proteases, by presenting them with a so-called bait-site, in the form of an arginine–threonine bond that they cleave. When C1r and C1s attack the bait-site they covalently bind C1INH and dissociate from C1q. By this mechanism, the C1 inhibitor limits the time during which antibody-bound C1 can cleave C4 and C2 to generate C4b,2b, the classical pathway C3 convertase.

Activation of C1 also occurs spontaneously at low levels without binding to an antigen:antibody complex, and can be triggered further by plasmin, a protease of the clotting system, which is also normally inhibited by C1INH. In the

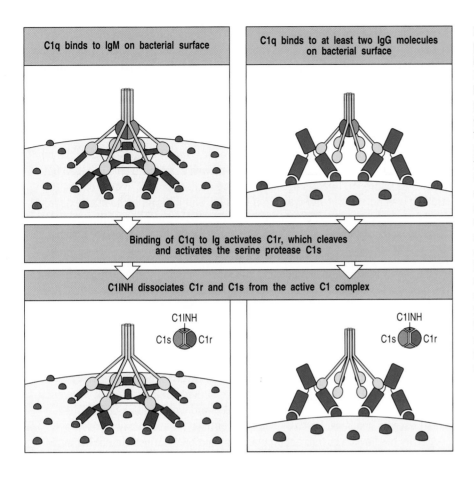

Fig. 9.2 Activation of the classical pathway of complement and intervention by C1INH. In the left panel, one molecule of IgM, bent into the 'staple' conformation by binding several identical epitopes on a pathogen surface, allows binding by the globular heads of C1q to its Fc pieces on the surface of the pathogen. In the right panel, multiple molecules of IgG bound to the surface of the pathogen allow binding by C1q to two or more Fc pieces. In both cases, binding of C1q activates the associated C1r, which becomes an active enzyme that cleaves the proenzyme C1s, a serine protease that initiates the classical complement cascade. Active C1 is inactivated by C1INH, which binds covalently to C1r and C1s, causing them to dissociate from the complex. There are in fact two C1r and two C1s molecules bound to each C1q molecule, although for simplicity this is not shown here. It takes four molecules of C1INH to inactivate all the C1r and C1s.

absence of C1INH, further active complement components are produced. This is seen in hereditary angioneurotic edema (HANE), a disease caused by a genetic deficiency of C1INH.

The case of Richard Crafton: a failure of communication as well as of complement regulation.

Richard Crafton was a 17-year-old high-school senior when he had an attack of severe abdominal pain at the end of a school day. The pain came as frequent sharp spasms and he began to vomit. After 3 hours, the pain became unbearable and he went to the emergency room at the local hospital.

At the hospital, the intern who examined him found no abnormalities other than dry mucous membranes of the mouth, and a tender abdomen. There was no point tenderness to indicate appendicitis. Richard continued to vomit every 5 minutes and said the pain was getting worse.

A surgeon was summoned. He agreed with the intern that Richard had an acute abdominal condition but was uncertain of the diagnosis. Blood tests showed an elevated red blood cell count, indicating dehydration. The surgeon decided to proceed with exploratory abdominal surgery. A large midline incision revealed a moderately swollen and pale jejunum but no other abnormalities were noted. The surgeon removed Richard's appendix, which was normal, and Richard recovered and returned to school 5 days later.

Richard, age 17, presents as an acute abdominal emergency.

Appendectomy performed. Appendix appears normal.

Family history of colic.

What Richard had not mentioned to the intern or to the surgeon was that, although he had never had such severe pains as those he was experiencing when he went to the emergency room, he had had episodes of abdominal pain since he was 14 years old. No one in the emergency room asked him if he was taking any medication or took a family history or a history of prior illness. If they had, they would have learned that Richard's mother, a maternal uncle, and maternal grandmother also had recurrent episodes of severe abdominal pain, as did his only sibling, a 19-year-old sister.

As a newborn, Richard was prone to severe colic. When he was 4 years old, a bump on his head led to abnormal swelling. When he was 7, a blow with a baseball bat caused his entire left forearm to swell to twice its normal size. In both cases, the swelling was not painful, nor was it red or itchy, and it disappeared after 2 days. At age 14 years, he began to complain of abdominal pain every few months, sometimes accompanied by vomiting and, more rarely, by clear, watery diarrhea.

Richard's mother had taken him at age 4 years to an immunologist, who listened to the the family history and immediately suspected hereditary angioneurotic edema. The diagnosis was confirmed on measuring key complement components. C1INH levels were 16% of the normal mean and C4 levels were markedly decreased, while C3 levels were normal.

When Richard turned up for a routine visit to his immunologist a few weeks after his surgical misadventure, the immunologist, noticing Richard's large abdominal scar, asked what had happened. When Richard explained, he prescribed daily doses of Winstrol (stanozolol). This caused a marked diminution in the frequency and severity of Richard's symptoms. When Richard was 20 years old, purified C1INH became available; he has since been infused intravenously on several occasions to alleviate severe abdominal pain, and once for swelling of his uvula, pharynx, and larynx. The infusion relieved his symptoms within 25 minutes.

Richard subsequently married and had two children. The C1INH level was found to be normal in both newborns.

Hereditary angioneurotic edema.

Individuals like Richard with a hereditary deficiency of C1INH are subject to recurrent episodes of circumscribed swelling of the skin (Fig. 9.3), intestine, and airway. When the swelling occurs in the intestine it causes severe abdominal pain, and obstructs the intestine so that the patient vomits. When the colon is affected, severe watery diarrhea may occur. Swelling in the larynx is the most dangerous symptom, as the patient may rapidly choke to death. Such episodes may be triggered by trauma, menstrual periods, excessive exercise, exposure to extremes of temperature, or mental stress. These types of event are associated with activation of four serine proteases, which are normally inhibited by C1INH. At the top of this cascade is Factor XII, which directly or indirectly activates the other three (Fig. 9.4). Factor XII is normally activated by injury to blood vessels, and initiates the kinin cascade, activating kallikrein, which generates the vasoactive peptide bradykinin. Factor XII also indirectly activates plasmin, which, as mentioned before, activates C1 itself. Plasmin also cleaves C2a to generate a vasoactive fragment called C2 kinin. In patients deficient in C1INH, the uninhibited activation of Factor XII leads to the activation of kallikrein and plasmin; kallikrein then catalyzes the formation of bradykinin, and plasmin activates C1, which cleaves C2, generating a small fragment, C2a, which is further cleaved by plasmin to produce C2 kinin. Bradykinin and C2 kinin increase the permeability of the

Fig. 9.3 Hereditary angioneurotic edema. Transient localized swelling that occurs in this condition often affects the face.

postcapillary venules by causing contraction of endothelial cells so as to create gaps in the blood vessel wall (Fig. 9.5). This is responsible for the edema; movement of fluid from the vascular space into another body compartment, such as the gut, causes the symptoms of dehydration as the vascular volume contracts.

Discussion and questions.

1 *Activation of the complement system results in the release of histamine and chemokines, which normally produce pain, heat, and itching. Why is the edema fluid in HANE free of cellular components, and why does the swelling not itch?*

Histamine release on complement activation is caused by C3a (the small cleavage fragment of C3) and the main chemokine is C5a (the small cleavage fragment of C5). These are both generated by the C3/C5 convertase, which in the classical pathway is formed from C4b and C2b. In HANE, C4b and C2b are both generated free in plasma. C4b is rapidly inactivated if it does not bind immediately to a cell surface; for that reason, and because the concentrations of C4b and C2b are relatively low, no C3/C5 convertase is formed, C3 and C5 are not cleaved and C3a and C5a are not generated.

The edema in HANE is caused not by the potent inflammatory mediators of the late events in complement activation but by C2a generated during the early events, and by bradykinin generated through the uninhibited activation of the kinin system.

2 *Richard has a markedly decreased amount of C4 in his blood. This is because it is being rapidly cleaved by activated C1. What other complement component would you expect to find decreased? Would you expect the alternative pathway components to be low, normal or elevated? What about the terminal components?*

The only other complement component that should be decreased is C2, which is also cleaved by C1. C1 plays no part in the alternative pathway of complement activation, so complement activation by the alternative pathway is not affected. The terminal components are not affected either. The unregulated activation of the early complement components does not lead to the formation of the C3/C5 convertase (see Question 1 above), so the terminal components are not abnormally activated. The depletion of the early components of the classical pathway does not affect the response to the normal activation of complement by bound antibody because the amplification of the response through the alternative pathway compensates for the deficiency in C4 and C2.

3 *Despite the complement deficiency in patients with HANE, they are not unduly susceptible to infection. Why not?*

This is not hard to explain: as we have already remarked, the alternative pathway of complement activation is intact and thus, although the classical pathway is affected by deficiencies in C2 and C4, these are compensated for by the potent amplification step from the alternative pathway.

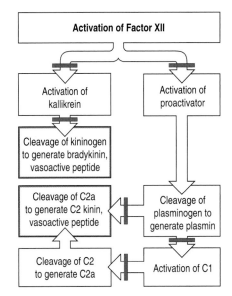

Fig. 9.4 Pathogenesis of hereditary angioneurotic edema. Activation of Factor XII leads to the activation of kallikrein, which cleaves kininogen to produce the vasoactive peptide bradykinin; it also leads to the activation of plasmin, which in turn activates C1. C1 cleaves C2, whose smaller fragment C2a is further cleaved by plasmin to generate the vasoactive peptide C2 kinin. The red bars represent inhibition by C1INH.

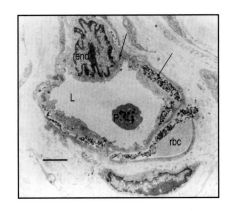

Fig. 9.5 Contraction of endothelial cells creates gaps in the blood vessel wall. A guinea pig was injected intravenously with India ink (a suspension of carbon particles). Immediately thereafter the guinea pig was injected intradermally with a small amount of activated C1s. An area of angiodema formed about the injected site, which was biopsied 10 minutes later. An electron micrograph reveals that the endothelial cells in post-capillary venules have contracted and formed gaps through which the india ink particles have leaked from the blood vessel. Micrograph courtesy of Kaethe Willms.

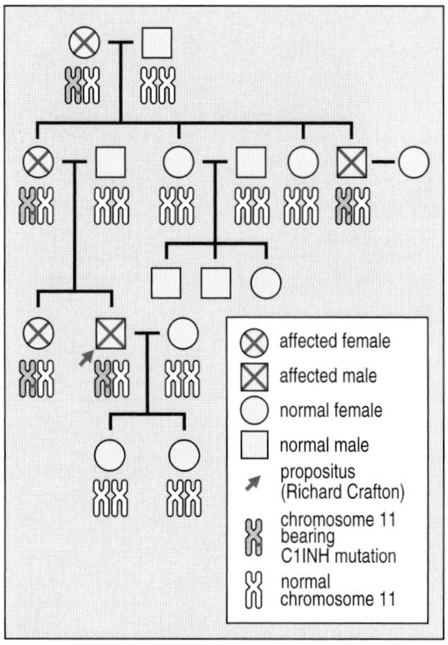
affected female
affected male
normal female
normal male
propositus (Richard Crafton)
chromosome 11 bearing C1INH mutation
normal chromosome 11

Fig. 9.6 The inheritance of HANE in Richard's extended family.

4 What is stanozolol, and why was it prescribed?

Stanozolol is a well-known anabolic androgen that has been used illegally by Olympic competitors. For unknown reasons, anabolic androgens suppress the symptoms of HANE, and that is why stanozolol was prescribed to Richard. Patients, especially females, do not like to take these compounds because they cause weight gain, acne, and sometimes amenorrhea. Preparations of purified C1INH are now available and intravenous injection of C1INH prepared from human donors is safe and very effective in halting the symptoms of the disease.

5 Emergency treatment for HANE cases is sometimes necessary because of airway obstruction. In most cases, however, a patient with obstruction of the upper airways is likely to be suffering from an anaphylactic reaction. The treatment in this case would be epinephrine. How might you decide whether to administer epinephrine or intravenous C1INH?

In practice, you would administer epinephrine immediately in any case, because most such emergencies are due to anaphylactic reactions and epinephrine is a harmless drug. If the laryngeal edema is anaphylactic, it will respond to the epinephrine. If it is due to hereditary angioneurotic edema, it will not. Anaphylactic edema is also likely to be accompanied by urticaria and itching, and the patient may have been exposed to a known allergen. Most patients know if they are allergic or have a hereditary disease, and they should be asked if they have had a similar problem before.

6 Fig. 9.6 shows Richard's family tree. What is the mode of inheritance (dominant or recessive, sex-linked or not) of HANE? Can Richard's two children pass the disease onto their offspring?

HANE does not skip generations: it is therefore likely that its effects are dominant. It clearly affects both males and females, so it cannot be sex-linked. If the gene has a dominant phenotype, and Richard's two children are normal, then it follows they cannot have inherited the defective gene from their father, and their children cannot inherit the disease from them.

Richard has inherited his abnormal C1INH gene from his mother. Since he has a normal C1INH gene from his father, you might expect that he would have 50% of the normal level of C1INH. However, the tests performed by his immunologist revealed 16% of the normal level. In general, functional C1INH tests in HANE patients reveals between 5% and 30% of normal activity. How could this be explained? There are two possibilities: decreased synthesis (that is, less than 50% synthesis from only one gene); or increased consumption of C1INH as a result of increased C1 activation. Both explanations have been shown to be correct. Patients with HANE synthesize about 37–40% of the normal amount of C1INH, and C1INH catabolism is 50% greater than in normal controls.

CASE 10 | Factor I Deficiency

> ## The alternative pathway of complement activation is important in innate immunity.

The complement system plays a crucial part in the destruction and removal of microorganisms from the body. Pathogens coated with complement proteins are more efficiently phagocytosed by macrophages, and bacteria coated with complement can also be directly destroyed by complement-mediated lysis. The system of plasma proteins known collectively as complement can be activated in various ways (see Fig. 9.1), of which the so-called alternative pathway is important in innate or non-adaptive immunity. This pathway can be activated in the absence of antibody, although even low titers of IgM antibodies against an infecting microorganism will greatly amplify complement activation.

The complement protein C3 is the starting point of the alternative pathway. It is one of the more abundant globulins in blood and is continuously being cleaved at a fairly low rate into a smaller C3a and a larger C3b fragment by a variety of host or microbial proteinases—this is called the 'tickover' of C3. Cleavage exposes a highly reactive thioester bond in the C3b fragment, which enables C3b to bond covalently with the hydroxyl group of serine or threonine in a protein or the hydroxyl group of a sugar on a microbial surface. If C3b fails to attach to a microbial surface, the thioester bond is spontaneously hydrolyzed and the C3b is inactivated (Fig. 10.1).

Topics bearing on this case:
Alternative pathway of complement activation
Factor I cleavage of C3b
Opsonizing activity of C3b
C3a activation of mast cells

Fig. 10.1 Cleavage of C3 exposes a reactive thioester bond that enables the larger cleavage fragment C3b to bind covalently to the bacterial cell surface. Intact C3 has a shielded thioester bond that is exposed when C3 is cleaved by a proteinase to produce C3b. The highly reactive thioester bond in C3b can react with hydroxyl or amino groups to form a covalent linkage with molecules on the microbial surface. In the absence of such a reaction the thioester bond is rapidly hydrolyzed, inactivating C3b.

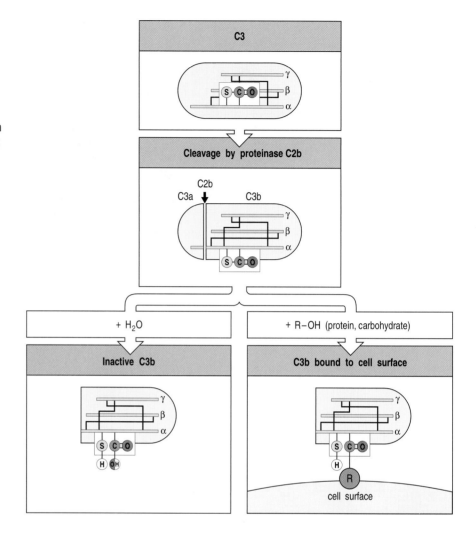

The binding of C3b to a microbial surface stimulates the cleavage of more C3 molecules. Another alternative pathway component, factor B, binds to C3b, and in this bound state is cleaved by a pre-existing blood proteinase, factor D, leaving the larger Bb fragment still bound to the C3b. The resulting C3b,Bb complex is an active serine protease, known as the alternative pathway C3 convertase, which specifically cleaves native C3 to make more C3b and C3a (Fig. 10.2).

Fig. 10.2 The alternative pathway of complement activation leads to amplification of C3 cleavage. C3b deposited on a microbial surface can bind factor B, making it susceptible to cleavage by factor D. The C3b,Bb complex is the C3 convertase of the alternative pathway of complement activation and its action results in the deposition of many more molecules of C3b on the pathogen surface.

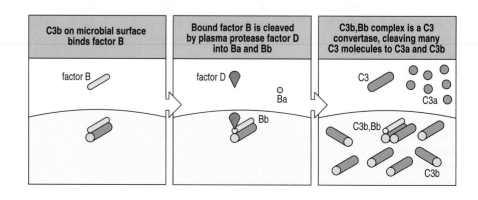

Fig. 10.3 The conversion of C3b to C3bi by factor I. On microbial surfaces, factor H displaces B from the C3b,Bb complex and cleaves C3b to produce C3bi. On host cell surfaces, the complement receptor CR1, which binds C3b, can substitute for factor H in this reaction.

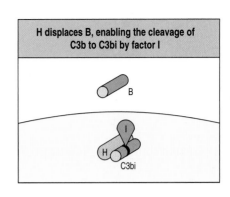

C3b bound to a microbial surface acts as an opsonin by binding to a specific receptor, the complement receptor 3 (CR3), on phagocytes, facilitating the ingestion of C3b-coated particles. But before C3b can act as a ligand for CR3, and thus as an effective opsonin, it has to undergo a further cleavage to a fragment called C3bi, which is effected by a blood serine protease called factor I, acting in conjunction with the blood protein factor H, components of the alternative pathway (Fig. 10.3). C3bi acting at the receptor CR3 can activate neutrophils and macrophages in the absence of antibody (Fig. 10.4). Cleavage of C3b by Factor I also has another critical effect. It inhibits the C3 convertase activity of the C3b complex, thus ensuring that supplies of C3 do not become depleted. On host cell surfaces, complement receptor 1 (CR1) may bind to C3b and serve as the co-factor instead of Factor H.

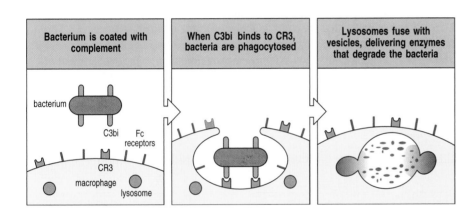

Fig. 10.4 C3bi binds to complement CR3 receptors on phagocytes and stimulates phagocytosis. C3bi on the surface of the pathogen binds to CR3 receptors on the phagocyte. Binding signals the phagocyte to engulf the particle and activates the internal destruction mechanisms.

The case of Morris Townsend: uncontrolled complement activation leads to susceptibility to infection and to hives.

Morris Townsend was admitted to the Brighton City Hospital at age 25 with pneumonia. This was his twenty-eighth admission to the hospital in his lifetime. From his first year onwards he had been repeatedly admitted for middle ear infections and mastoiditis. During these episodes, which were successfully treated with antibiotics, a variety of pyogenic (pus-forming) bacteria were cultured from his ears or mastoids, including *Staphylococcus aureus*, *Proteus vulgaris* and *Pseudomonas aeruginosa*. At age 3 he had a tonsillectomy and adenoidectomy because of enlargement and chronic infection of his nasopharyngeal lymphoid tissue; at age 6 he had scarlet fever. He had also been admitted at other times with left lower lobe pneumonia (when *Haemophilus influenzae* had been cultured from his sputum), an abscess in the groin, acute sinusitis, a posterior ear abscess due to *Corynebacterium* species, skin abscesses with accompanying bloodstream infection (septicemia) due to β-hemolytic streptococci and, on one occasion, septicemia due to *Neisseria meningitidis* (meningococcemia).

25-year-old male with repeated bacterial infections. Complement or Ig deficiency?

On physical examination at his latest admission, Morris was found to be slightly obese but otherwise normally developed. His hearing was poor in both ears and this was attributed to his recurrent ear infections and mastoiditis. He also told doctors that he developed hives all over his body after drinking alcohol or after taking a bath or shower.

A urine analysis yielded normal results. His hematocrit was 43% (normal) and his white cell count was 6000 μl^{-1}. His platelet count was 240,000 μl^{-1} (normal) and his blood clotted normally. His red blood cells gave a strong positive agglutination reaction with an antibody to C3 but no agglutination with an antibody to IgG or IgM. His serum IgG level was 915 mg dl^{-1}, IgA 475 mg dl^{-1} and IgM 135 mg dl^{-1} (all normal). Morris responded normally to an injection of tetanus toxoid; his antibody titer rose from 0.25 to 8.0 hemagglutinating units ml^{-1}. He gave a positive delayed-type skin reaction to mumps and monilia antigens.

Serum levels of C3 were 27 mg dl^{-1} (normal values 97–204 mg dl^{-1}); of this 8 mg dl^{-1} was C3 and 19 mg dl^{-1} C3b. The serum levels of all other complement components were normal except for factor B, which was undetectable. His serum failed to kill a smooth strain of *Salmonella newport*, even after addition of C3 to the serum to render the C3 concentration normal. To investigate the turnover of C3, Morris was injected with a dose of C3 labeled with the radioactive tracer ^{125}I. The results of this investigation showed that the rate of synthesis of C3 was normal but that C3 was being broken down at four times the normal rate (Fig. 10.5). A test of his serum with an antibody to factor I showed that his serum lacked factor I.

Morris's family had no history of recurrent bacterial infections, but investigations showed reduced levels of factor I in both his parents and in several of his siblings (Fig. 10.6).

Ig normal; check C3.

C3 synthesis normal; check factor I.

Genetic deficiency? Test family for factor I.

Factor I deficiency.

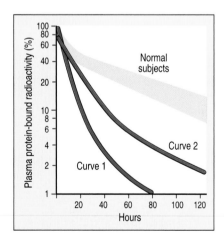

Fig. 10.5 The rate of disappearance of ^{125}I-labeled C3 from the plasma. The rate of disappearance of radioactively labeled C3 from the patient's plasma (curve 1) is much faster than that in normal subjects (shaded area; 11 normal subjects). Curve 2 shows the rate of disappearance of C3 in the patient after the infusion of 500 ml of normal plasma.

Patients such as Morris Townsend with a genetic deficiency in factor I were instrumental in deciphering the mechanism of activation of the alternative pathway of complement. Innate immunity is a first and highly effective means of defense against the common extracellular bacteria that cause pyogenic infections. The lack of factor I means that the alternative pathway C3 convertase is uninhibited and consumption of C3 is greatly accelerated, leading to C3 depletion. The lack of C3, and the non-production of C3bi, results in defective opsonization, which is the main means of removing and destroying these bacteria. Thus, factor I deficiency, like the genetic deficiency of C3 itself, results in a greatly increased susceptibility to infections with such bacteria. The clinical findings in factor I deficiency are not unlike those observed in X-linked agammaglobulinemia—a failure of opsonization results in frequent pyogenic infections.

The gene encoding factor I is on chromosome 4. The family of Morris Townsend provides a classic case of the inheritance of a recessive mutation in an autosomal gene (see Fig. 10.6). His parents, some of his siblings and one nephew are heterozygous for the defect; they produce roughly half the normal amounts of factor I, which is sufficient to prevent any clinical symptoms. Morris appears to be the only family member who is homozygous for the defect, and who thus exhibits symptoms.

Two interesting facts emerge from his clinical history. He sustained recurrent hives and had one bout of meningococcemia. It is easy to understand why he had hives. He was constantly cleaving C3 to C3a and C3b. C3a binds to mast cells and, among other things, causes the release of histamine and hence hives. The interesting question is why he did not have hives all the time and why they became problematic only after the ingestion of alcohol or exposure to hot and cold water. We must suppose that he had tachyphylaxis, or end-organ unresponsiveness, to histamine. Only when histamine release was increased, as by alcohol consumption or sudden changes in ambient body temperature, did the symptoms appear.

Morris Townsend's meningococcemia in particular is symptomatic of a deficiency in components of the alternative pathway of complement action. There are two common human pathogens in the bacterial genus *Neisseria*: *Neisseria gonorrhoeae* and *Neisseria meningitidis*. The former causes the sexually transmitted disease gonorrhea; the latter causes septicemia and meningitis and can be rapidly fatal. Patients have died from septic shock within 20 minutes of the onset of the symptoms of meningococcemia. Patients with genetic defects in the alternative pathway of complement activation or in the terminal components of complement sustain overwhelming and repeated infection with *Neisseria*. Deficiencies in the alternative pathway components factor D and properdin (factor B deficiency has never been observed in humans) were discovered because these patients developed recurrent meningococcemia. Similar observations have been made in patients with deficiencies of the later-acting C5, C6, C7, C8, and C9 components. These clinical observations highlight the importance of the bactericidal action of complement in controlling septicemia due to *Neisseria*.

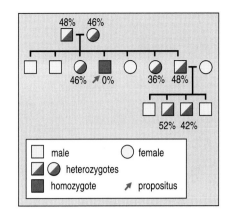

Fig. 10.6 Autosomal recessive inheritance of the factor I defect. The numbers indicate percentages of normal level of factor I in serum. No number indicates normal levels. The blue arrow indicates the patient, Morris Townsend.

Discussion and questions.

$\boxed{1}$ *Morris Townsend's clinical course has improved with age and he now has far fewer infections than he had as a child and adolescent. How do you explain this?*

As he lives longer his adaptive immunity against these common bacteria becomes better and he has come to rely less on innate immune mechanisms for protection against infection. Bacteria coated with antibodies can be phagocytosed independently of complement via the Fc receptors on phagocytes.

$\boxed{2}$ *From the radiolabeled C3 experiment, we found that Morris Townsend catabolized the C3 very quickly but that his rate of synthesis of C3 was normal. What do you anticipate would happen if we repeated the experiment with radiolabeled factor B?*

A similar result. The factor B would be rapidly destroyed and its rate of synthesis could turn out to be normal. The overproduction of C3b in the absence of factor I leads to an increased binding of factor B to C3b and its subsequent cleavage by factor D. Thus, factor B is being consumed excessively as a result of the deficiency in factor I. The lack of factor B, like the C3 deficiency, is secondary to the basic defect in factor I.

3   *Morris Townsend was given a large dose of pure factor I intravenously. What changes would you predict to occur in his serum proteins?*

His serum levels of C3 and factor B rose to normal. C3b disappeared from his serum. The effect lasted for about 10 days.

4   *What other genetic defect in the alternative pathway might lead to the same clinical and laboratory results as factor I deficiency?*

Deficiency in factor H. Because factor H is needed for the cleavage of C3b by factor I in the blood, factor H deficiency should result in clinical symptoms identical with those of factor I deficiency. In fact this is true; several families with factor H deficiency have been studied and they show symptoms indistinguishable from Factor I deficiency.

5   *Why did Morris' red blood cells agglutinate with antibody to C3?*

He is producing large amounts of C3b, which binds to complement receptor 1 (CR1) on red blood cells and leads to their agglutination by anti-C3.

CASE 11 | AIDS in Mother and Child

Infection can suppress adaptive immunity.

Certain infectious microorganisms can suppress or subvert the immune system. For example, in lepromatous leprosy, *Mycobacterium leprae* induces T cells to produce lymphokines that stimulate a humoral response but suppress the development of a successful inflammatory response to contain the leprosy bacillus. The leprosy bacillus multiplies and there is a persistent depression of cell-mediated immune responses to a wide range of antigens (Case 29). Another example of immunosuppression is provided by bacterial super-antigens, such as toxic shock syndrome toxin-1. Superantigens bind and stimulate large numbers of T cells by binding to certain V_β chains of the T-cell receptor, inducing massive production of cytokines by the responding T cells (Case 7). This, in turn, causes a temporary suppression of adaptive immunity.

Topics bearing on this case:
Failure of cell-mediated immunity
Infection with the human immuno-deficiency virus (HIV)
Control of HIV infection
Drug therapy for HIV infection
ELISA test

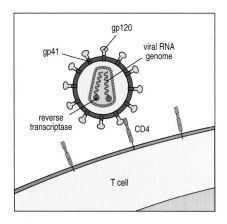

Fig 11.1 HIV binds to CD4 T cells through its coat glycoprotein gp120. The gp120 molecule on the surface of the virus binds CD4 on T cells and macrophages; the viral protein gp41 then mediates fusion of the enveloped virus with the target cell, allowing the viral genome to enter the cell.

At the beginning of this century, when tuberculosis was the leading cause of death and fully half the population was tuberculin-positive, it was well known that an intercurrent measles infection would cause a well-contained tuberculosis infection to run rampant and result in death. The mechanism responsible is now known to be the suppression of IL-2 synthesis after binding of measles virus to CD46 on macrophages.

Some of the microorganisms that suppress immunity act by infecting lymphocytes. Infectious mononucleosis or glandular fever is caused by a virus (Epstein–Barr virus) that infects B lymphocytes. The infection activates cytotoxic CD8 T cells, which destroy the B cells in which the Epstein–Barr virus is replicating. In the third week of infection, at the height of activation of CD8 T cells, all adaptive immunity is suppressed. The lymphokines responsible for the immunosuppression are not well defined but probably include interleukin-10 and TGF-β (Case 12).

The human immunodeficiency virus (HIV) presents a chilling example of the consequences of infection and destruction of immune cells by a microorganism. The T-cell surface CD4 molecule acts as the receptor for HIV (Fig. 11.1). CD4 is also expressed on the surface of cells of the macrophage lineage and they too can be infected by this virus. The chemokine receptors CCR5 and CXCR4 act as obligatory co-receptors for HIV. As we shall see, the primary infection with HIV may go unnoticed, and the virus may replicate in the host for many years before symptoms of immunodeficiency can be seen. During this period of clinical latency, the level of virus in the blood and the number of circulating CD4 cells remains fairly steady but in fact both virus particles and CD4 cells are being rapidly destroyed and replenished, as rounds of virus replication take place in newly infected cells. When the rate at which CD4 cells are being destroyed exceeds the capacity of the host to replenish them, their number decreases to a point where cell-mediated immunity falters. As we have seen in other cases, such as severe combined immunodeficiency (see Case 5), the failure of cell-mediated immunity renders the host susceptible to fatal opportunistic infections.

The Pinkerton family: a tragedy from contaminated blood.

Six-month-old infant with thrush, diarrhea and otitis media.

Benjamin Pinkerton was a captain in the United States Navy, stationed in Japan. He married a Japanese woman before leaving Japan for his new assignment in Honolulu, Hawaii. His wife, Chieko, gave birth to a healthy daughter in 1987. Two years later they had a son, Franklin, who weighed 8 pounds at birth and appeared to be very healthy. At age 3, 4, and 5 months, Franklin received routine immunization with tetanus and diphtheria toxoids and pertussis bacteria (DPT) as well as oral polio vaccine. He had no reactions to these inoculations and seemed to be thriving. At 6 months of age he became sick and lost weight. He developed severe, persistent diarrhea with fever. He was noted to have white spots (thrush) in his mouth; he had infections of the middle ear (otitis media) twice in rapid succession. Franklin was seen several times at the Naval Hospital by pediatricians who treated him with antibiotics but he did not seem to get better. At 7 months of age Franklin developed mild difficulty in breathing.

Fig. 11.2 Use of the enzyme-linked immunosorbent assay (ELISA) to detect the presence of antibodies to the HIV coat protein gp120. Purified recombinant gp120 is coated onto the surface of plastic wells to which the protein binds non-specifically; residual sticky sites on the plastic are blocked by adding irrelevant proteins (not shown). Serum samples from the individuals being tested are then added to the wells under conditions where non-specific binding is prevented, so that only binding to gp120 causes antibodies to be retained on the surface. Unbound antibody is removed from all wells by washing, and anti-human immunoglobulin that has been chemically linked to an enzyme is added, again under conditions that favour specific binding alone. After further washing, the colourless substrate of the enzyme is added, and colored material is deposited in the wells in which the enzyme-linked anti-human immunoglobulin is found. This assay allows arrays of wells known as microtiter plates to be read in fiberoptic multichannel spectrometers, greatly speeding the assay.

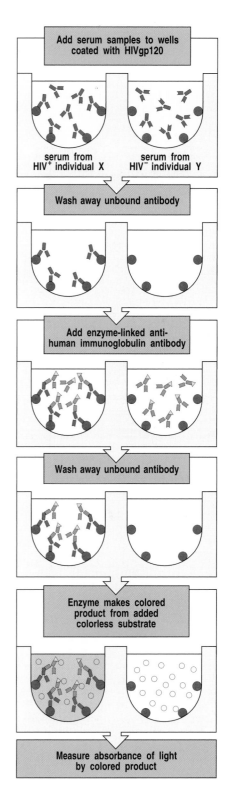

Add serum samples to wells coated with HIVgp120

serum from HIV+ individual X   serum from HIV− individual Y

Wash away unbound antibody

Add enzyme-linked anti-human immunoglobulin antibody

Wash away unbound antibody

Enzyme makes colored product from added colorless substrate

Measure absorbance of light by colored product

On physical examination at this time his temperature was recorded at 38°C (normal 37°C). Thrush was evident as white plaques (*Candida* spp.) on the buccal mucosa (the inside of his mouth) and anterior surface of his tongue (see Fig. 5.4). A diaper rash was also present that seemed to be due to *Candida* infection. He had fine inspiratory rales (crackles) in both lungs. He was admitted to the hospital, where his white blood cell count was found to be normal at 6500 μl^{-1} with a normal differential count of 62% neutrophils, 5% monocytes, 30% lymphocytes, 2% eosinophils and 1% basophils. His serum immunoglobulin G measured 997 mg dl^{-1} (normal), IgM 73 mg dl^{-1} (normal), and IgA 187 mg dl^{-1} (normal). An examination of his lymphocytes revealed 1825 CD8 T cells μl^{-1} (normal) but only 85 CD4 T cells μl^{-1} (very depressed). He showed no delayed-type hypersensitivity (DTH) response to intradermal *Candida* antigen or to PPD (purified protein derivative of tuberculosis). His serum contained antibodies to HIV by ELISA testing (Fig. 11.2) and Western blot (Fig. 11.3). After this had been discovered, his mother, father and sister were also tested by the same methods. His mother and father tested positive for antibodies to HIV. His sister tested negative.

Although Captain Pinkerton was in good health, his wife, Chieko, had felt run down, and complained of low-grade fevers and swollen lymph nodes in her neck. She attributed all these symptoms to the stress of Franklin's illness. However, it turned out that a year before Franklin's birth, Chieko had been pregnant. Near the end of her pregnancy the fetus died and had to be removed by Caesarian section. She recovered from the Caesarian section surgery and felt perfectly well, but because of blood loss during the surgery, she had been given 2 units of blood.

In the hospital, bronchial washings were obtained from Franklin. A stain for *Pneumocystis carinii* was positive and he was treated with the drugs trimethoprim and sulfa, which cleared the infection. Nonetheless, Franklin did not reach his developmental milestones on schedule. He was unable to crawl at 10 months. He became unsteady when sitting. He lost 1 kg of body weight. Despite the successful eradication of his *Pneumocystis carinii* infection, Franklin's breathing became more rapid and his lungs appeared worse on radiographs. Culture of tissue from an open lung biopsy grew cytomegalovirus, respiratory syncytial virus, and *Pseudomonas aeruginosa*. A duodenal biopsy also contained cytomegalovirus. He developed a severe cough and was spitting up blood (hemoptysis). One week later he died of respiratory failure.

Chieko, in the meanwhile, was started on AZT (zidovudine) therapy to combat her HIV infection but she developed *Pneumocystis carinii* pneumonia when her CD4 T-cell number fell below 200 μl^{-1}. Despite successful treatment with trimethoprim and sulfa, the infection recurred. She died 5 months after Franklin from respiratory failure. Captain Pinkerton has remained asymptomatic despite the persistence of HIV antibodies in his serum.

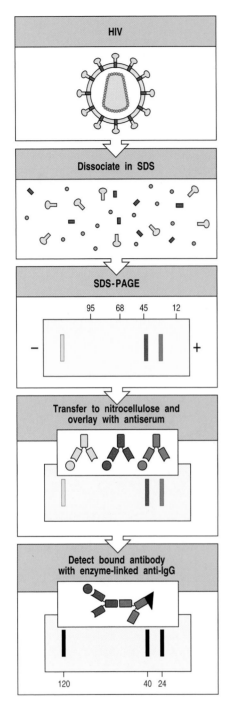

HIV

Dissociate in SDS

SDS-PAGE

95   68   45   12

–   +

Transfer to nitrocellulose and overlay with antiserum

Detect bound antibody with enzyme-linked anti-IgG

120   40 24

Fig. 11.3 Western blotting is used to identify antibodies to the human immunodeficiency virus (HIV) in serum from infected individuals. The virus is dissociated into its constituent proteins by treatment with the detergent SDS, and its proteins are separated by SDS-PAGE. The separated proteins are transferred to a nitrocellulose sheet and reacted with the test serum. Anti-HIV antibodies in the serum bind to the various HIV proteins and are detected by using enzyme-linked anti-human immunoglobulin, which deposits colored material from a colorless substrate. This general methodology will detect any combination of antibody and antigen and is used widely, although the denaturing effect of SDS means that the technique works most reliably with antibodies that recognize the antigen when it is denatured.

Acquired Immune Deficiency Syndrome (AIDS).

AIDS is caused by the human immune deficiency virus (HIV) of which there are two known types, HIV-1 and HIV-2. HIV-2 was largely confined to West Africa but now seems to be spreading into Southeast Asia. HIV infections in North and South America and in Europe are exclusively from HIV-1. HIV can be transmitted by homosexual and heterosexual intercourse, by infusion of contaminated blood or blood products, or by contaminated needles, which are the major source of infection among drug addicts. The infection can also be passed from mother to child during pregnancy, during delivery or, more uncommonly, by breastfeeding. Somewhere between 25% and 35% of infants born to HIV-positive mothers are infected. The rate of infection of infants has been decreased four-fold by giving HIV-positive pregnant women the anti-viral agent AZT (zidovudine).

Contact with the virus does not necessarily result in infection. The standard indicator of infection is the presence of antibodies to the virus coat protein gp120. The initial infection, as in Chieko's case, may pass unnoticed and without symptoms. More often a mild viral illness within 6 weeks of infection is sustained, with fever, swollen lymph nodes and a rash. It subsides at about the time that seroconversion occurs, and although virus and antibody persist, the patient feels well. A period of clinical latency lasting years, and perhaps even decades, may ensue during which the infected person feels perfectly well. Then he/she begins to experience low-grade fever and night sweats, excessive fatigue, and perhaps oral candidiasis (thrush) in the mouth. Lymph nodes in the neck or axillae or groin may swell. Weight loss may become very marked. These are the prodromal symptoms of impending AIDS. (A prodrome is a concatenation of signs and symptoms that predict the onset of a syndrome.) The number of CD4 T cells in the blood may have been normal up to this time but, with the onset of the prodrome, the CD4 T cell count begins to fall (Fig. 11.4). When the number of CD4 T cells decreases to the range of 200–400 cells μl^{-1}, the final phase of the illness, which is called AIDS, starts. At this time serious, eventually fatal, opportunisitic infections as well as certain unusual malignancies occur (Fig. 11.5).

At any time after the infection, HIV may infect megakaryocytes, which have some surface CD4. Because megakaryocytes are the bone-marrow progenitors of blood platelets, extensive infection of megakaryocytes causes the platelet count to fall (thrombocytopenia) and bleeding to occur. HIV may also infect the glial cells of the brain. Glial cells are of monocyte–macrophage lineage and have some CD4 on their surface. The infection of glial cells may cause dementia and other neurological symptoms, such as the motor problems and developmental retardation seen in Franklin.

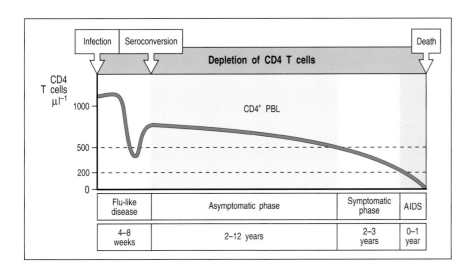

Fig. 11.4 The typical course of infection with HIV. The first few weeks are typified by an acute influenza-like viral illness, sometimes called seroconversion disease, with high titers of virus in the blood. An adaptive immune response follows, which controls the acute illness and largely restores CD4 T cell levels but does not eradicate the virus. Opportunistic infections and other symptoms become more frequent as the CD4 T-cell count falls, starting at around 500 cells μl^{-1}. The disease then enters the symptomatic phase.

Discussion and questions.

1 *When Franklin was seen by the pediatricians at the Naval Hospital, their first impression was that he had severe combined immunodeficiency (SCID) because of the cough, thrush and persistent diarrhea. What laboratory findings directed their attention to the diagnosis of AIDS?*

His serum immunoglobulin levels were normal. In SCID they would be very low. His T cells showed a specific deficit of CD4$^+$ T cells. In SCID the total number of T cells would be decreased.

2 *If a lymph-node biopsy had been obtained from Franklin, in what way would it differ from the histopathology of a lymph node if he had SCID or X-linked agammaglobulinemia (XLA)?*

The lymph node would be enlarged and exhibit marked, if not exuberant, follicular hyperplasia. In SCID and XLA the lymph nodes are very small. In SCID, the node would contain no or very, very few lymphoid cells. In XLA, the lymph node would have no follicles, no germinal centers, and no B cells or plasma cells. T cells would be present but not in an organized array (Fig. 11.6).

Fig. 11.5 A variety of opportunistic pathogens and cancers can kill AIDS patients. Infections are the major cause of death in AIDS, with respiratory infection with *Pneumocystis carinii* being the most prominent. Most of these pathogens require effective macrophage activation by CD4 T cells or effective cytotoxic T cells for host defense. Opportunistic pathogens are present in the normal environment but cause severe disease primarily in immuno-compromised hosts, such as AIDS patients and cancer patients. AIDS patients are also susceptible to several rare cancers, such as Kaposi's sarcoma and lymphomas, suggesting that immune surveillance by T cells may normally prevent such tumors. EBV, Epstein–Bar virus.

Infections	
Parasites	*Toxoplasma* spp. *Cryptosporidium* spp. *Leishmania* spp. *Microsporidium* spp.
Bacteria	*Mycobacterium tuberculosis* *Mycobacterium avium* *intracellulare* *Salmonella* spp.
Fungi	*Pneumocystis carinii* *Cryptococcus neoformans* *Candida* spp. *Histoplasma capsulatum* *Coccidioides immitis*
Viruses	Herpes simplex Cytomegalovirus Varicella zoster

Malignancies
Kaposi's sarcoma (invasive) Non-Hodgkin's lymphoma, including   EBV-positive Burkitt's lymphoma Primary lymphoma of the brain

Fig. 11.6 Lymph node sections from patients with SCID, XLA, and AIDS.

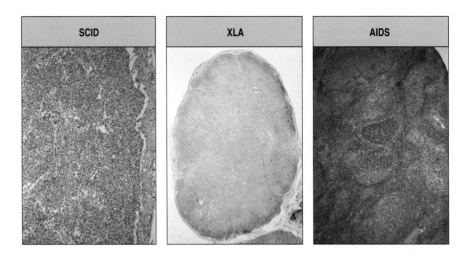

| SCID | XLA | AIDS |

3 *The course of the HIV infection was very different in Franklin and in his mother and illustrates the differences between pediatric AIDS and adult AIDS. What is the major difference and how do you account for it?*

In infants, the HIV infection typically runs a more rapid course. Franklin died before he reached 1 year of age. His mother was infected but without symptoms for 4 years. The difference in the course of AIDS in infants and in adults is probably due to the fact that the infection occurs in an immunologically immature and naive subject when it affects infants, whereas an infected adult has a functionally mature immune system and decades of acquired adaptive immunity. This has consequences for both the response to HIV itself, and the susceptibility to other infections. We have seen that newborn T cells are not fully 'turned on'. For example, in the hyper IgM case (Case 3), we saw that the CD40 ligand is not readily expressed on activation of the T cells of newborns. Their T cells do not synthesize interferon-γ in normal amounts. Their cytotoxic T lymphocytes are not readily activated. This functional immaturity is probably why young infants have difficulty confining and walling off infections, particularly those that require adaptive immunity mediated by T cells. Tuberculosis offers a clear example of this. It is a fast-spreading, highly lethal infection in young infants, whereas in older children and adults who are immunologically normal, this infection is usually confined to the lung, or more rarely to other organs. HIV infection in infants occurs before they have had an opportunity to develop any adaptive immunity to common infections, and this means they are prone to certain infections not seen in adult AIDS. Thus an adult will already have antibodies to the common pyogenic bacteria and will not be particularly susceptible to pyogenic infections, whereas these are frequently observed in affected infants. Adults will also have been exposed to common viruses. Epstein–Barr Virus (EBV), for example, is normally encountered early in life and contained as a latent infection. Virtually all adults have been infected with EBV by the end of the second or third decade of life, and primary EBV infection is therefore not a threat to HIV-infected adults. For an HIV-infected infant, however, a first encounter with this virus causes bizarre manifestations such as parotitis (inflammation of the parotid gland, like mumps) and a form of pneumonia characterized by pulmonary lymphoid hyperplasia (Franklin did not have these, or any other evidence of a primary EBV infection).

---

4 *What are the mechanisms of resistance to the progression of HIV infection?*

The answer to this question is not known precisely. The immune response to the virus is illustrated in Fig. 11.7. Antibody to HIV seems to play a minor role in resisting the progress of infection. HIV-specific cytotoxic CD8 cells arise as virus levels decline from the peak associated with primary infection, and seem to have a more important role in containing the infection. Rare individuals with mutations in the co-receptors for HIV (CCR5 and CXCR4) are resistant to HIV infection.

5 *A few individuals, mostly hemophiliacs, are known to have been infected with HIV as long as 18 years ago and yet they remain asymptomatic. What factors may contribute to long-term survival with this infection?*

The virus burden in these individuals is very low and, in some cases, the only detectable viruses carry mutations in genes such as *nef* or *tat*, which are vital to HIV replication in the infected host. However, viruses able to replicate in culture can be isolated from most of these so-called 'long-term non-progressors.' These patients seem to contain replication-competent virus, most probably by continuing to maintain a successful cytotoxic CD8 T-cell response.

6 *What is the mode of action of AZT or zidovudine?*

AZT (3′-azido,2′,3′-dideoxythymidine) is a nucleotide analog that is phosphorylated inside the cell, and used as a substrate by the reverse transcriptase of HIV (Fig. 11.8). HIV reverse transcriptase synthesizes a DNA complement of the viral RNA, at the start of a new round of virus replication in a newly infected cell. The incorporation of AZT blocks further extension of this DNA strand and thereby stops replication of the virus. Two other nucleotide analogs, ddI (dideoxyinosine) and ddC (dideoxycytosine) inhibit HIV replication by a similar mechanism and are also used to treat HIV infection. Unfortunately, mutation allows the virus to acquire resistance to all these drugs. Replication of HIV (and other known retroviruses) is error-prone, and the virus mutates as it replicates in the infected host. Resistance to AZT requires multiple mutations but can arise in only a few months. Combining HIV protease inhibitors with zidovudine or other nucleotide analogs has dramatically improved the survival of HIV-infected patients.

Fig. 11.7 The immune response to HIV. Infectious virus is present at relatively low levels in the peripheral blood of infected individuals during a prolonged asymptomatic phase but is replicated persistently in lymphoid tissues. During this period, CD4 T-cell counts gradually decline, although antibodies and CD8 cytotoxic T cells directed against the virus remain at high levels. Two different antibody responses are shown in the figure, one to the envelope protein of HIV, env, and one to the core protein p24. Eventually, the levels of antibody and HIV-specific cytotoxic T lymphocytes (CTLs) also fall, and progressively more infectious HIV appears in the peripheral blood.

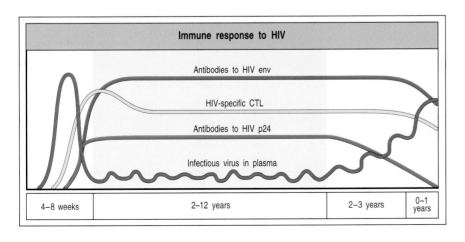

Immune response to HIV

Antibodies to HIV env

HIV-specific CTL

Antibodies to HIV p24

Infectious virus in plasma

| 4–8 weeks | 2–12 years | 2–3 years | 0–1 years |

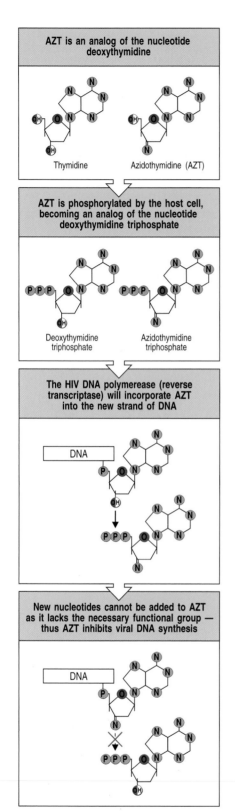

Fig. 11.8 The mechanism of action of the AIDS drug AZT.

7 What is the mechanism of CD4 T-cell depletion in HIV infection?

The precise answer to this question is also unknown. It is clear that the cells producing the virus are killed, either by cytotoxic T lymphocytes or by direct cytotoxic effects of the virus. It is also possible that the death of non-infected 'bystander' cells contributes to CD4 T-cell depletion. HIV is known to have a cytotoxic effect on CD4 T cells in culture. The viral capsular gp120 binds and crosslinks the CD4 molecule, which depresses T-cell function and may induce apoptosis, even when the CD4 cells are not themselves infected by HIV. If, as seems likely, the early killing of HIV-infected CD4 cells by cytotoxic T cells serves to contain the virus and prevent greater CD4 cell depletion in the next round of HIV infection and replication, a declining ability to mount cytotoxic responses, especially to new viral variants, could be very important. Patients infected with HIV show impairment of T-cell function, especially memory cell responses, even during the asymptomatic phase. They are hypergammaglobulinemic, and humoral responses seem favored at the expense of cell-mediated immunity. This immunoregulatory bias impedes inflammatory responses, and possibly also cytotoxic responses to HIV-infected cells.

8 What is the most important known determinant of the progression of HIV infection?

The CD4 T-cell count is, statistically speaking, the best indicator of the time-course of progression to AIDS. Other factors such as lifestyle and the incidence of intercurrent infections do not seem statistically significant.

9 Which cytokine, released in HIV infection, causes weight loss?

Tumor necrosis factor-alpha (TNF-α).

10 What steps have been taken since the early 1980s to prevent a tragedy such as occurred in the Pinkerton family? What pitfalls are encountered in screening blood and blood products?

All blood banks screen blood donations for antibody to HIV gp120 by enzyme-linked immunosorbent assay (ELISA) and confirm positive results with Western blotting. The major pitfall in relying on this approach is its inability to detect an HIV-infected blood donor in the period between acquiring HIV and the formation of antibody to HIV. Such individuals can be detected only by testing blood for HIV by polymerase chain reaction (PCR).

CASE 12 | Acute Infectious Mononucleosis

Cytotoxic T cells terminate viral infection.

All viruses, and some bacteria, multiply inside infected cells; indeed, viruses are highly sophisticated parasites that do not have a complete biosynthetic or metabolic apparatus of their own and, in consequence, must replicate inside a living cell. Once inside a cell, a pathogen is not accessible to antibodies and has to be eliminated by other means.

Some intracellular bacteria live and multiply in membrane-bound phagosomes within macrophages and are killed by antibacterial agents released into these vacuoles after macrophage activation by CD4 T_H1 cells (see for example Case 29). Viruses, in contrast, together with those bacteria that live in the cytosol, can be eliminated only by destruction of the infected cell itself. This role in host defense is fulfilled by the cytotoxic CD8 T cells of adaptive immunity and the natural killer (NK) cells of innate immunity.

CD8 cytotoxic T cells kill infected cells by recognizing foreign, pathogen-derived peptides that are transported to the cell surface bound to MHC class I molecules (see Fig. 4.1). The peptides carried by MHC class I molecules come from the degradation of proteins in the cytosol and so cytotoxic T cells act against pathogens whose proteins are found in the cytosol of the host cell

Topics bearing on this case:

Activation of cytotoxic T cells

Cell killing by cytotoxic T cells

Processing and presentation of cytosolic antigens

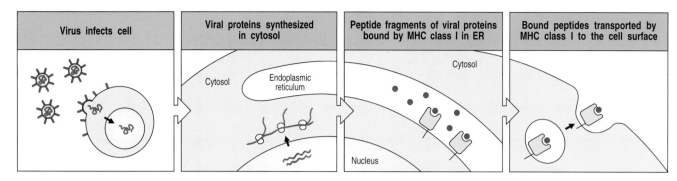

Fig. 12.1 MHC class I molecules present antigen derived from proteins in the cytosol. In cells infected with viruses, viral proteins are synthesized in the cytosol. Peptide fragments of viral proteins are transported into the endoplasmic reticulum, where they are bound by MHC class I molecules, which then deliver the peptides to the cell surface.

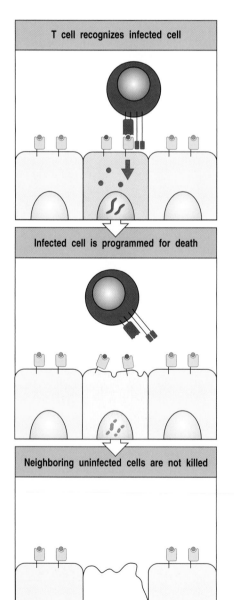

at some stage in their life-cycle (Fig. 12.1). The critical role of cytotoxic CD8 T cells in host defense is seen in the increased susceptibility of animals artificially depleted of cytotoxic T cells to many viral and intracytosolic bacterial infections. Mice and humans lacking the MHC class I molecules that present antigen to CD8 cells are also more susceptible to such infections.

Cytotoxic T cells kill their infected targets with great precision and neatness, by inducing apoptosis in the infected cell while sparing adjacent normal cells; this strategy minimizes tissue damage (Fig. 12.2). CD8 cytotoxic T cells release two types of pre-formed cytotoxin—the fragmentins, which seem able to induce apoptosis in any type of target cell, and the pore-forming protein perforin, which punches holes in the membrane of the target cell through which the fragmentins can enter (Fig. 12.3). A membrane-bound molecule, the Fas ligand, that is expressed on CD8 T cells as well as on some CD4 T cells can also induce apoptosis by binding to Fas on a limited range of target cells. Together, these properties allow the cytotoxic T cell to attack and destroy virtually any infected cell. Cytotoxic CD8 T cells also produce the cytokine interferon (IFN)-γ; this cytokine inhibits viral replication, induces MHC class I expression, and also activates macrophages. As well as combatting infection by viruses and intracytosolic bacteria, CD8 T cells are important in controlling some protozoal infections; they are crucial, for example, in host defense against *Toxoplasma gondii*, an intracellular protozoan.

The importance of cytotoxic T cells in the control of viral replication is highlighted by many aspects of Epstein–Barr virus (EBV) infection, which is described in this case study. EBV (also known as human herpesvirus 4) is a member of the virus family Herpetoviridae. It has a double-stranded linear DNA genome enclosed in an icosahedral capsid and a lipid envelope and replicates its DNA genome in the host cell nucleus. EBV infects only humans and is one of the most successful infective agents on its obligate host. It can even be thought of as a commensal that only seldom causes injury to the host; anywhere from 60–98% of healthy adults show serological evidence of infection with EBV. The virus infects mainly B cells and epithelial cells.

Fig. 12.2 Cytotoxic T cells kill target cells bearing specific antigen while sparing neighboring uninfected cells. All the cells in a tissue are susceptible to the induction of apoptosis by the cytotoxins of armed effector CD8 T cells but only infected cells are killed. Specific recognition by the T-cell receptor identifies which target cell to kill, and the polarized release of cytotoxic granules (see Fig. 12.3) ensures that neighboring cells are spared.

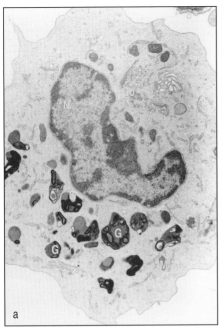
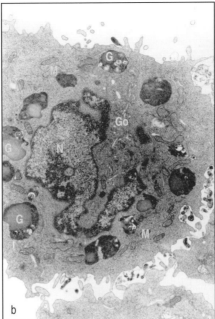
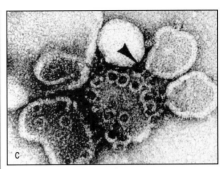
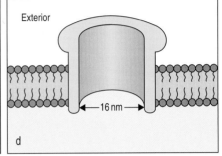

Fig. 12.3 Perforin released from the granules of cytotoxic T cells can insert into the target-cell membrane to form pores. Perforin molecules, as well as several other effector molecules, are contained in the granules of cytotoxic T cells (panel a: G = granules; N = nucleus; M = mitochondria; Go = Golgi apparatus). When a CD8 cytotoxic T cell recognizes its target, the granules are released onto the target cell (panel b, bottom right quadrant). The perforin molecules released from the granules polymerize and insert into the membrane of the target cells to form pores. The structure of these pores is best visualized when purified perforin is added to synthetic lipid vesicles (panel c: pores are seen both end on, as circles, and sideways on, arrow). The pores span the target cell membrane (panel d). Photographs courtesy of E Podack.

The case of Emma Bovary: a bad sore throat from a B-cell infection.

Emma Bovary was a healthy 15-year-old when she suddenly developed a very sore throat accompanied by fever and malaise. Her throat was so swollen she had difficulty swallowing. Over the next few days the fever waxed and waned, her sore throat became worse and she became progressively more tired and anorectic (lost her appetite). On the third day of illness her pediatrician noted severe pharyngitis and took a throat culture for β-hemolytic streptococci; the culture proved negative.

Emma's symptoms persisted, and she was unable to eat as she could hardly swallow. She said she had no difficulty breathing but that her left upper abdomen felt slightly uncomfortable. Emma's 1-year-old brother became ill at the same time, but did not have such severe symptoms. He was merely listless and felt warm. He had no particular physical symptoms, and seemed to recover completely after a few days.

On physical examination on the tenth day of illness, Emma appeared very ill. She had a high temperature (38.2 °C), pulse rate of 84, respiratory rate of 18, and blood pressure 85/55. Her mouth was dry and her tonsils were red and enlarged. They met in the midline, leaving a passage of only 2 × 2 cm approximately. Palatal petechiae

(very small hemorrhages under the mucosa) could be seen. Her anterior and posterior cervical lymph nodes were swollen and tender (lymphadenopathy); the largest nodes were 2 × 2 cm. Her abdomen felt soft and the liver was enlarged, the edge being palpable 2 cm below the right costal margin. The spleen was also enlarged; the tip was easily palpable under the left costal margin.

A blood test gave a white blood cell count of 18,590 μl^{-1} with 39% neutrophils, 27% lymphocytes, 22% atypical lymphocytes (very high), and 11% monocytes (high); her hematocrit was 45% and the platelet count 397,000 μl^{-1}. Serum electrolytes were normal. Another throat culture was obtained and blood tests for Epstein–Barr virus (EBV) were ordered.

In the meantime a presumptive diagnosis of acute infectious mononucleosis was made with complications including partial pharyngeal obstruction and mild dehydration. Emma was admitted to the hospital and received 1 liter normal saline intravenously followed by 20 mg methylprednisolone (a corticosteroid) intravenously every 12 hours.

Her throat culture again proved negative for streptococcus but her blood serum was positive for IgM and IgG antibodies against EBV capsid antigen. Emma improved quickly with the symptomatic treatment and was discharged on the second day after admission to complete her recovery at home.

Acute infectious mononucleosis.

Emma shows many of the clinical features characteristic of acute infectious mononucleosis (IM) induced by the Epstein–Barr virus, a disease also known as glandular fever in some countries. She had severe pharyngitis with petechiae on the palate, swollen lymph nodes in the neck, enlarged liver and spleen, and large numbers of atypical lymphocytes (the mononucleosis after which the disease is named) in her blood. These cells, also known as Downey–McKinlay cells, are large cells with foamy basophilic cytoplasm and fenestrated nuclei. They are mostly T cells, with a preponderance of CD8 cytotoxic T cells, and are present in 90% of patients with IM, where they sometimes constitute the majority of blood leukocytes.

It is these cells that control the acute infection by destroying EBV-infected B cells. The atypical lymphocytosis in IM is a reflection of the increased CD8 T cell cytotoxic activity. In the vast majority of individuals, the infection is brought under control but not eradicated, because the viral genome persists latently in many B cells. In the latent phase some viral antigens are produced and peptides derived from them are presented by MHC I molecules at the surface of the infected cell, thus enabling latently infected cells to be recognized and destroyed by EBV-specific cytotoxic T cells. Some of the latently infected B lymphocytes become transformed; they are able to propagate themselves indefinitely if removed from the presence of EBV-specific cytotoxic T cells, and are potentially malignant. In healthy people after infection, approximately 1 in 10^6 B cells is transformed. In patients with immune deficiency or who are immunosuppressed, infected B cells can grow unchecked. In immuno-deficient patients, EBV infection can cause immunoblastic lymphoma and T-cell lymphoma.

EBV has a long incubation period: the time between primary EBV infection and the onset of illness is 30–50 days. Infection in infancy or early childhood is almost always asymptomatic, or results in mild disease, as evidenced by Emma's younger brother, who probably also had a primary EBV infection. In developed countries, primary infection is delayed in about half of the population to adolescence and early adulthood, so that 30–50% of primary EBV infection results in acute IM.

EBV enters B cells by binding to the B-cell surface molecule CD21 (also called complement receptor 2 (CR2) because it acts as a receptor for the complement fragment C3dg). This receptor is also present on a small subpopulation of T cells and on various types of epithelial cell in the nasopharynx, parotid gland duct, female cervix, and male urethra. After an active phase of viral multiplication, latency is established in B cells and, in some cases, epithelial cells. The EBV DNA is maintained during latency as an extrachromosomal DNA within the nucleus. Virus production is reactivated from time to time and periodic shedding of infectious virus in oral secretions of healthy infected people is common and lifelong. Adolescents like Emma often catch the disease through kissing; although Emma was not yet sexually active, she had dated several boys in her class.

Definitive diagnosis of EBV infection is best made by serological or molecular biological tests. Infected B cells are stimulated to secrete immunoglobulin, producing, among other antibodies, a so-called heterophile IgM antibody whose detection is one of the most widely used diagnostic tests for EBV infection. This heterophile antibody is not specific for EBV antigens but binds to antigens present on heterologous red blood cells (that is, those of other animals, such as sheep or goat) and agglutinates them. In addition, specific antibody responses are generated against several EBV-specific antigens, and the appearance of different antibodies is informative as to the time of infection and the pattern of virus replication (Fig. 12.4). The presence of IgM antibody against the EBV capsid antigen (VCA) indicates that the infection is acute; this antibody declines gradually in the convalescent phase. Because of the long incubation period of the virus, anti-VCA IgG antibody is also detectable at the onset of illness. The continued presence of EBV antigens in the host maintains antibody production throughout life.

Antibodies to so-called early antigens (EA) are produced mainly in the convalescent phase (1–6 months after disease onset); they then disappear. The absence of EA antibody indicates that the virus is mostly quiescent, and is not undergoing replication on a large scale. If the infection is reactivated,

	Viral capsid antigen; IgM	Viral capsid antigen; IgG	Early antigen	Epstein–Barr virus nuclear antigen
Never exposed	–	–	–	–
Acute infection	+	+	+/–	–
Recent infection	+/–	+	+/–	+/–
Past infection	–	+	–	+
Reactivated or chronic infection	–	+	+/–	+/–

Fig. 12.4 Serologic diagnosis of EBV infection.

EA antibody titers rise again. Antigens expressed later in the viral life cycle, during its latent phase, are the EBNAs (Epstein–Barr nuclear antigens). Appearance of EBNA antibody indicates that the virus has been present in the body for at least a few months.

In very young, immunosuppressed, or immunodeficient patients, antibody formation can be so impaired that serologic diagnosis is impossible. In these cases EBV antigens can be detected by immunofluorescence microscopy on blood or tissue (e.g. lymph node) specimens or by *in situ* hybridization for small EBV RNAs (EBERs). Alternatively, viral DNA can be amplified from infected cells and tissue by the polymerase chain reaction with oligo-nucleotide primers specific for EBV DNA.

EBV infection is mitogenic for B cells, over-riding the normal regulatory mechanisms preventing them from dividing. Activation of B cells by virus infection also leads to immunoglobulin production, as noted above. Several EBV genes are critical for B-cell activation. One is the viral gene *BCRF-1,* which encodes a protein, also called VIL-10, very similar to human interleukin-10 (IL-10). The BCRF-1 protein enhances the activation and proliferation of EBV-infected cells. EBV infection also stimulates endogenous synthesis of IL-10 and IL-6, with further autocrine stimulatory effects on B cells. IL-10 also inhibits the T-cell production of IL-2 and IFN-γ, and enhances the production of B-cell stimulatory cytokines such as IL-4. IL-6 may inhibit the ability of NK cells to destroy EBV-infected cells.

In most patients IM is a self-limited disease for which supportive therapy suffices. Nucleoside analogs such as acyclovir or ganciclovir, or the DNA polymerase inhibitor foscarnet, have limited ability to inhibit the replication of EBV *in vitro*. The clinical usefulness of these drugs is so far unproven. They are frequently administered to patients with fulminant disease, or to immunosuppressed patients. Corticosteroids are often prescribed as a palliative measure, especially when airway obstruction is a potential concern. In the most extreme cases, when respiratory distress is present, tonsillectomy can be required. Corticosteroids reduce virus shedding and provide some symptomatic relief due to their anti-inflammatory effects. They do not significantly alter the course of the disease.

The ability of EBV to transform B cells is an extremely useful laboratory tool. When peripheral blood B cells are cultured with EBV, they become immortalized at a relatively high frequency and can be propagated indefinitely *in vitro*. This allows the general study of various aspects of B-cell biology, as well as providing material for the study of individual patients.

The virus-encoded Epstein–Barr virus nuclear antigen 2 (EBNA-2) is critical for transformation. EBNA-2 protein interacts with transcription factors leading to the activation of several host genes such as those encoding B-cell activating molecules such as CD21 (the EBV receptor) and CD23 (FcεRII, the low-affinity receptor for IgE), and viral genes such as LMP-1 (latent membrane protein-1, the primary oncogene of EBV). EBNA-3, -4, -5, and -6 also have a role in B-cell transformation.

Acute IM is only one of the possible outcomes of EBV infection. The Epstein–Barr virus is in fact named after two workers who studied Burkitt's lymphoma in Africa in the 1960s and first cultured the virus from these patients. This B-cell lymphoma is strongly associated with EBV infection in Africa, but not in other parts of the world; the high rate of EBV infection in early infancy together with the high incidence of malaria in Africa seem to be the predisposing factors. EBV infection is also strongly associated with nasopharyngeal carcinoma in Southeast Asia. This may perhaps be due to a particular strain of EBV that no longer possesses the epitope provoking an immunodominant response in people with a certain HLA class I allele that is present in a high proportion of people in Southeast Asia. Such people are therefore not so efficient at clearing cells infected with this EBV strain, making transformation and eventual malignancy more likely.

Discussion and questions.

---

1 *Patients with humoral immunodeficiency (an impaired antibody response) are susceptible to infection with some viruses such as poliomyelitis or enteric viruses but they have no problems with EBV. Why?*

Cell-mediated cytotoxic activity against EBV-infected cells is the main method of controlling EBV replication. Although EBV-specific antibodies are produced in a normal infected host, they do not seem to have a major role in controlling the virus. Patients who lack the ability to produce specific antibody, but who have intact T-cell cytotoxic responses, are in most instances able to fight the infection effectively.

2 *There is a high risk of EBV-induced lymphoproliferative malignancy after T-cell depleted bone marrow transplantation. Why?*

Bone marrow transplantation is most often performed in patients whose immune systems have either been destroyed with high doses of chemotherapy or in patients with primary immune deficiency. Even when donor and recipient are matched for the major histocompatibility antigens, some degree of minor incompatibility is inevitable. A major complication that can arise after a bone marrow transplant is graft-versus-host disease (GVHD). In GVHD, mature T lymphocytes in the bone marrow graft are activated and begin to attack the host's tissues. To prevent this, bone marrow grafts are often treated to remove mature T cells. If a bone marrow donor has been infected with EBV, they will have a small number of transformed B cells carrying the virus (about one in a million B cells). In the donor, these B cells are being 'held in check' by cytotoxic cells; there is an equilibrium between cell division and death of EBV-infected B cells. If mature T cells are removed from a marrow graft, but B cells are left, then the transformed B cells will escape from surveillance by cytotoxic T cells and might begin to proliferate at a high rate. Removing both mature B and T cells from donor bone marrow results in a lower rate of EBV lymphoproliferative disease after transplantation. An alternative explanation is that if a transplant recipient is infected with EBV the destruction of their immune system before transplantation removes enough cytotoxic cells to tip the balance in favor of the donor transformed B cells.

Treatment of EBV-related lymphoproliferative disease after transplantation has demonstrated the potential utility of adoptive immunotherapy. EBV-specific cytotoxic T cells are used directly from donor blood or can be expanded *in vitro* by culture with EBV-transformed donor cells. The cytotoxic T cells are then transfused into the transplant recipient with lymphoproliferative disease and can kill the dividing donor B cells. This treatment has achieved

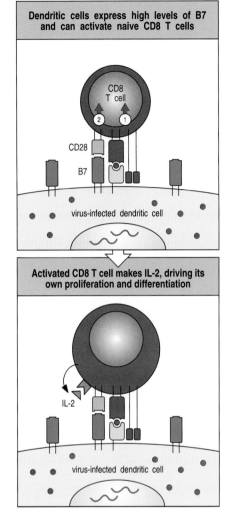

Dendritic cells express high levels of B7 and can activate naive CD8 T cells

CD8 T cell

CD28

B7

virus-infected dendritic cell

Activated CD8 T cell makes IL-2, driving its own proliferation and differentiation

IL-2

virus-infected dendritic cell

Fig. 12.5 Naive CD8 T cells can be activated directly by potent antigen-presenting cells. Naive CD8 T cells encountering peptide:MHC class I complexes on the surface of dendritic cells that express high levels of co-stimulatory molecules (top panel) are activated via separate signals (1 and 2) to produce IL-2 (bottom panel) and proliferate in response to it. They eventually differentiate into armed CD8 cytotoxic T cells that are then able to attack any cell-type infected with the same virus. Interaction of CD40L on helper T cells and CD40 on dendritic cells strongly enhances the capacity of dendritic cells to activate CD8 cytotoxic T cells, by causing an increased expression of co-stimulatory molecules and secretion of cytokines, eg IL-12.

remission of EBV lymphoproliferative disease in up to 90% of patients in various studies so far. The major complication of this treatment is acute or chronic GVHD.

3 In vitro *transformation by EBV of B cells from umbilical cord blood rarely fails. Transformation of B cells in blood cultures from some adults is difficult. Why?*

Many adults will have circulating in their blood activated and armed cytotoxic T cells specific for EBV (Fig. 12.5). When a blood sample is cultured and then infected with EBV, the B cells will become infected and the activated cytotoxic T cells will then immediately destroy the infected cells. Transformed B cells can be prepared from these individuals either by removing T cells from the blood sample before culture or by adding inhibitors of T-cell activation such as cyclosporin A. It is exceedingly unlikely that a fetus will have been infected by EBV, and so their blood sample is unlikely to contain any activated EBV-specific cytotoxic T cells.

4 *Why is heterophile antibody produced during EBV infection?*

EBV activates B cells polyclonally, that is without respect to the antigen specificities of the infected cells. A significant percentage (5–10%) of circulating B lymphocytes bear antigen receptors of low affinity for several cross-reacting carbohydrate, nucleotide, or glycoprotein antigens. If infected with EBV and activated, these cells will begin to secrete polyspecific IgM antibodies. Many of these antibodies will bind relatively non-specifically to erythrocytes of other species such as horse, ox, cow, or sheep and are thus called heterophile ('other-loving') antibodies. They are found in approximately 90% of patients with EBV.

5 *Males with X-linked agammaglobulinemia (XLA) never get infected with EBV. How do you explain this?*

These patients have no B cells, the cellular host of EBV.

6 *Males with X-linked lymphoproliferative disorder (X-LP) succumb to EBV-induced proliferation of B lymphocytes. Why?*

These boys suffer from mutations in the *X-LP* (*SAP*) locus on the X-chromosome. (SAP stands for SLAM-associated protein.) The X-LP/SAP gene product is expressed in T cells and interacts with the T-cell co-stimulatory molecule SLAM. This molecule has a role in T-cell production of IFN-γ, a cytokine that is important for the activity of cytotoxic T cells.

CASE 13 | Leukocyte Adhesion Deficiency

The traffic of white blood cells.

Newly differentiated blood cells continually enter the bloodstream from their sites of production: red blood cells, monocytes, granulocytes, and B lymphocytes from the bone marrow; and T lymphocytes from the thymus. Under ordinary circumstances red blood cells spend their entire life span of 120 days in the bloodstream. However, white blood cells are destined to emigrate from the blood to perform their effector functions. Lymphocytes recirculate through secondary lymphoid tissues, where they are detained if they encounter an antigen to which they can respond; macrophages migrate into the tissues as they mature from circulating monocytes; effector T lymphocytes and large numbers of granulocytes are recruited to extravascular sites in response to infection or injury. For example, it is estimated that, each day, three billion neutrophils enter the oral cavity, the most contaminated site in our body.

Topics bearing on this case:

Migration and homing of leukocytes

Phagocytic cell defects

Adhesion molecules in T-cell interactions

Bone marrow transplantation

Complement receptors

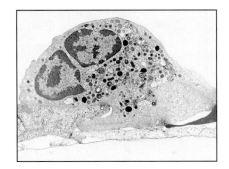

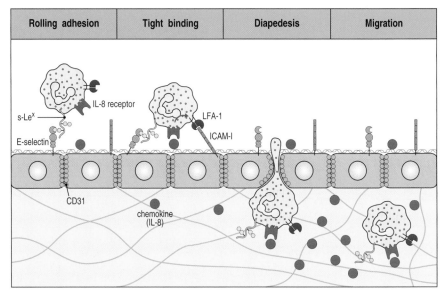

Fig. 13.1 Phagocytic leukocytes and effector T cells are directed to sites of infection through interactions between adhesion molecules induced by monokines. The first step (right panel) involves the reversible binding of leukocytes to vascular endothelium through interactions between selectins induced on the endothelium and their carbohydrate ligands on the leukocyte, shown here for E-selectin and its ligand the sialyl-Lewis x moiety (s-Le^x). This interaction cannot anchor the cells against the shearing force of the flow of blood and instead they roll along the endothelium, continually making and breaking contact. The binding does, however, allow stronger interactions, which occur as a result of the induction of ICAM-1 on the endothelium and the activation of its receptors LFA-1 and Mac-1 (not shown) on the leukocyte. This is mediated by chemokines (eg IL-8) retained on heparan sulfate proteoglycans on the endothelial cell surface. Tight binding between these molecules arrests the rolling and allows the leukocyte to squeeze between the endothelial cells forming the wall of the blood vessel (extravasate). The leukocyte integrins LFA-1 and Mac-1 are required for extravasation, and for migration toward chemoattractants. Adhesion between molecules of CD31, expressed on both the leukocyte and the junction of the endothelial cells, is also thought to contribute to diapedesis. Finally, the leukocyte migrates along a concentration gradient of chemokines (here shown as IL-8) secreted by cells at the site of infection. The electron micrograph shows a neutrophil that has just started to migrate between two endothelial cells (bottom of photo). Note the pseudopod that the neutrophil has inserted between adjacent endothelial cells. The dark mass at the bottom right is an erythrocyte that has become trapped underneath the neutrophil. Photograph ($\times$ 5500) courtesy of I Bird and J Spragg.

The process by which white blood cells migrate from the bloodstream to sites of infection is fairly well understood (Fig. 13.1). First their flow is retarded by the interaction between selectins whose expression is induced on activated vascular endothelium and certain fucosylated glycoproteins on the white cell surface (for example Sialyl-Lewis^x). Tight binding of leukocytes to the endothelial surface is then triggered by chemokines, such as interleukin-8, which activate an enhanced ability of the leukocyte integrins (for example LFA-1 and Mac-1) to adhere to their receptors. Crossing the endothelial cell wall also involves interactions between the leukocyte integrins and their receptors, while the subsequent direction of migration follows a concentration gradient of chemokines (for example IL-8) produced by cells already at the site of infection or injury. The process by which lymphocytes home to secondary lymphoid tissue is very similar, except that it is initiated by mucin-like addressins on lymphoid venules binding to L-selectin on the surface of naive lymphocytes (Fig. 13.2). The various adhesion molecules involved in leukocyte trafficking and other important leukocyte cell–cell interactions are described in Fig. 13.3.

An excellent opportunity to study the role of integrins is provided by a genetic defect in CD18, the common β chain of the three β$_2$ integrins: LFA-1 (CD11a/CD18), Mac-1 (CD11b/CD18, also known as CR3), and p150,95 (CD11c/CD18, also known as CR4). Children with this genetic defect suffer from leukocyte adhesion deficiency. They have recurrent pyogenic infections,

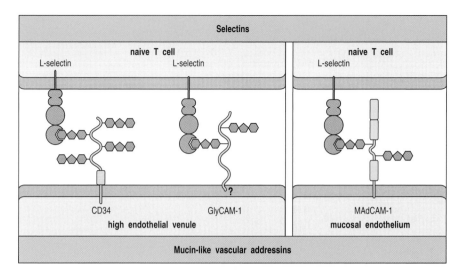

Fig. 13.2 L-selectin and the mucin-like vascular addressins direct naive lymphocyte homing to lymphoid tissues. L-selectin is expressed on naive T cells, which bind to the vascular addressins CD34 and GlyCAM-1 on high endothelial venules to enter lymph nodes. The relative importance of CD34 and GlyCAM-1 in this interaction is not clear. GlyCAM-1 is expressed exclusively on high endothelial venules but has no transmembrane region and it is unclear how it is attached to the membrane; CD34 has a transmembrane anchor and is expressed on endothelial cells but not exclusively on high endothelial venules. The addressin MAdCAM-1 is expressed on mucosal endothelium and guides entry into mucosal lymphoid tissue. L-selectin recognizes carbohydrate moieties on the vascular addressins.

problems with wound healing, and if they survive long enough they develop severe inflammation of the gums (gingivitis). Surprisingly, children with leukocyte adhesion deficiency are not unduly susceptible to opportunistic infections. This implies normal T-cell function despite the absence of LFA-1, which was thought to be important for T-cell adhesion to antigen-presenting cells (Fig. 13.4). The capacity to form antibodies is also unimpaired, showing that adequate collaboration between T and B cells can also occur without LFA-1.

The case of Luisa Ortega: a problem of immobile white blood cells.

Luisa Ortega was born at full term and weighed 3.7 kg. She was the second child born to the Ortegas. At 4 weeks of age Luisa was taken by her parents to her pediatrician because she had swelling and redness around the umbilical cord stump (omphalitis), and a fever of 39°C. Her white blood cell count was 71,000 μl^{-1} (normal 5000–10,000 μl^{-1}). She was treated in the hospital with intravenous antibiotics for 12 days and then discharged home with oral antibiotics. At the time of discharge her white blood count was 20,000 μl^{-1}. Cultures obtained from the inflamed skin about the umbilical stump before antibiotic treatment grew *Escherichia coli* and *Staphylococcus aureus*.

The Ortegas had a baby boy 3 years prior to Luisa's birth. At 2 weeks of age he developed a very severe infection of the large intestine (necrotizing enterocolitis). Separation of his umbilical cord was delayed. He subsequently suffered from multiple skin infections and he died of staphylococcal pneumonia at 1 year of age. Just before his death his white blood cell count was recorded at 75,000 μl^{-1}.

Because of the previous family history, Luisa was referred to the Children's Hospital. At the time of her admission to the Children's Hospital she seemed normal on physical examination, and radiographs of the chest and abdomen were normal.

Cultures of urine, blood and cerebrospinal fluid were negative. Her white blood count was 68,000 μl^{-1} (very elevated). Of her white cells, 73% were neutrophils, 22% lymphocytes, and 5% eosinophils (this distribution of cell types is in the normal range but the absolute count for each is abnormally high). Her serum IgG

Luisa, 4 weeks old, presents with acute omphalitis.

4-week-old infant with family history of overwhelming infection.

Fig. 13.3 Adhesion molecules in leukocyte interactions. Several structural families of adhesion molecules play a part in leukocyte migration, homing and cell–cell interactions: the selectins; mucin-like vascular addressins; the integrins; proteins of the immunoglobulin superfamily; and the protein CD44. The figure shows schematic representations of an example from each family, a list of other family members that participate in leukocyte interactions, their cellular distribution, and their partners in adhesive interactions. The nomenclature of the different molecules in these families is confusing because it often reflects the way in which the molecules were first identified rather than their related structural characteristics. Thus while all the ICAMs are immunoglobulin-related, and all the VLA molecules are β_1 integrins, the CD nomenclature reflects the characterization of leukocyte cell-surface molecules by raising monoclonal antibodies against them, and embraces adhesion molecules in all the structural families. The LFA molecules were defined through experiments in which cytotoxic T-cell killing could be blocked by monoclonal antibodies against cell-surface molecules on the interacting cells, and there are LFA molecules in both the integrin and the immunoglobulin families. Alternative names for each of the molecules shown are given in parentheses. Sialyl Lewis^x, which is recognized by P- and E-selectin, is an oligosaccharide present on cell-surface glycoproteins of circulating leukocytes.

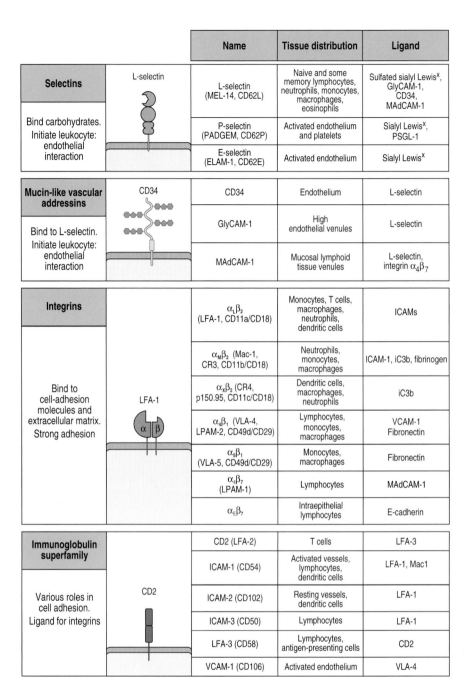

		Name	Tissue distribution	Ligand
Selectins Bind carbohydrates. Initiate leukocyte: endothelial interaction	L-selectin	L-selectin (MEL-14, CD62L)	Naive and some memory lymphocytes, neutrophils, monocytes, macrophages, eosinophils	Sulfated sialyl Lewis^x, GlyCAM-1, CD34, MAdCAM-1
		P-selectin (PADGEM, CD62P)	Activated endothelium and platelets	Sialyl Lewis^x, PSGL-1
		E-selectin (ELAM-1, CD62E)	Activated endothelium	Sialyl Lewis^x
Mucin-like vascular addressins Bind to L-selectin. Initiate leukocyte: endothelial interaction	CD34	CD34	Endothelium	L-selectin
		GlyCAM-1	High endothelial venules	L-selectin
		MAdCAM-1	Mucosal lymphoid tissue venules	L-selectin, integrin $\alpha_4\beta_7$
Integrins Bind to cell-adhesion molecules and extracellular matrix. Strong adhesion	LFA-1	$\alpha_L\beta_2$ (LFA-1, CD11a/CD18)	Monocytes, T cells, macrophages, neutrophils, dendritic cells	ICAMs
		$\alpha_M\beta_2$ (Mac-1, CR3, CD11b/CD18)	Neutrophils, monocytes, macrophages	ICAM-1, iC3b, fibrinogen
		$\alpha_x\beta_2$ (CR4, p150.95, CD11c/CD18)	Dendritic cells, macrophages, neutrophils	iC3b
		$\alpha_4\beta_1$ (VLA-4, LPAM-2, CD49d/CD29)	Lymphocytes, monocytes, macrophages	VCAM-1 Fibronectin
		$\alpha_5\beta_1$ (VLA-5, CD49d/CD29)	Monocytes, macrophages	Fibronectin
		$\alpha_4\beta_7$ (LPAM-1)	Lymphocytes	MAdCAM-1
		$\alpha_E\beta_7$	Intraepithelial lymphocytes	E-cadherin
Immunoglobulin superfamily Various roles in cell adhesion. Ligand for integrins	CD2	CD2 (LFA-2)	T cells	LFA-3
		ICAM-1 (CD54)	Activated vessels, lymphocytes, dendritic cells	LFA-1, Mac1
		ICAM-2 (CD102)	Resting vessels, dendritic cells	LFA-1
		ICAM-3 (CD50)	Lymphocytes	LFA-1
		LFA-3 (CD58)	Lymphocytes, antigen-presenting cells	CD2
		VCAM-1 (CD106)	Activated endothelium	VLA-4

concentration was 613 mg dl^{-1} (normal), her IgM was 89 mg dl^{-1} (normal), and her IgA 7 mg dl^{-1} (normal). The concentration of complement component C3 in her serum was 185 mg dl^{-1} and that of C4 was 28 mg dl^{-1} (both normal).

A Rebuck skin window was performed. In this procedure, the skin of the forearm is gently abraded with a scalpel blade and a cover slip is placed on the abrasion. After 2 hours the cover slip is removed and replaced by another every subsequent 2 hours for a total of 8 hours. In this way, the migration of immune cells into the damaged skin can be monitored. No white cells accumulated on the cover slips. All of Luisa's blood leukocytes, however, were present in abnormally high numbers. Of her blood lymphocytes 53% (7930 μl^{-1}) were T cells (as measured by CD3 expression); of these, 36% were CD4 and 16% CD8 (normal proportions); 25% (3740 μl^{-1}) were B cells (as measured by antibody to CD19), and 14% were NK cells (as measured by CD16 expression). These were both elevated.

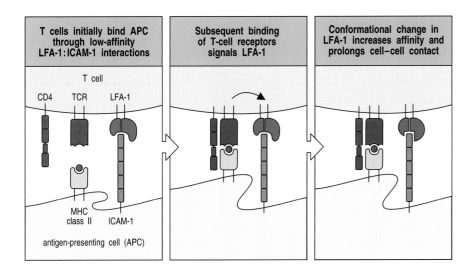

Fig. 13.4 Transient adhesive interactions between T cells and antigen-presenting cells are stabilized as specific antigen recognition induces high-affinity binding by LFA-1. When a T cell binds to its specific ligand on an antigen-presenting cell (APC), intracellular signaling through the T-cell receptor (TCR) induces a conformational change in LFA-1 that causes it to bind with higher affinity to ICAM-3 on the antigen-presenting cell. The cell shown here is a CD4 T cell.

Proliferation of Luisa's T cells in response to phytohemagglutinin (PHA) and concanavalin A (ConA) was slightly depressed. Further flow cytometric analysis revealed that, whereas 60% of Luisa's lymphocytes were stained by a monoclonal antibody to CD3, only 5% reacted with a monoclonal antibody to CD18, giving a CD18/CD3 ratio of 5%/60% compared with 62%/65% on testing cells from a control subject (Fig. 13.5). Her blood mononuclear cells were stimulated with PHA and examined after 3 days of incubation with a monoclonal antibody to CD11a (the α chain of LFA-1). No LFA-1 expression was found.

Luisa was treated with busulfan, cyclophosphamide and anti-thymocyte serum for 10 days. After this therapy, she was given bone marrow cells from her mother at a dose of 500×10^6 per kg body weight, and a short course of immunosuppressive therapy. Her mother's bone marrow donation had been depleted of mature T cells with a monoclonal antibody to mature T cells and complement. Twenty-eight days after the transplant, the lymphoid and myeloid cells in Luisa exhibited complete chimerism. She subsequently did well clinically and her white blood count remained at 7800 μl^{-1}.

Leukocyte adhesion deficiency.

Children like Luisa are subject to recurrent, severe bacterial infections that are eventually fatal. In these patients, encapsulated bacteria are coated with antibody and complement. However, the neutrophils and monocytes that would normally be recruited to the site of infection are entrapped in the bloodstream and cannot emigrate into the tissues because they lack LFA-1 (CD11a,CD18) and Mac1/CR3 (CD11b,CD18). In a normal individual, the first cover slip in a Rebuck glass window would contain many neutrophils. Monocytes begin to appear at 4 hours and by 8 hours the cover slip contains predominantly monocytic cells and very few neutrophils. In Luisa's case the cover slips had no cells because her leukocytes were unable to emigrate from the bloodstream and onto the cover slip. For this reason, the white cells in the bloodstream are very high: a very high white blood cell count is characteristic of leukocyte adhesion deficiency. The ability to deal with pyogenic bacteria is

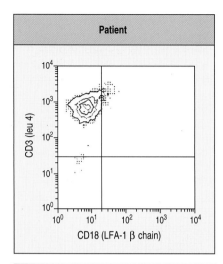

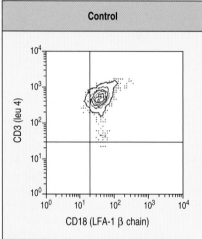

Fig. 13.5 Flow cytometric analysis of CD18 expression on Luisa's lymphocytes (top panel) compared with a control subject (bottom panel). The vertical axis measures fluorescence from labeled antibody to CD3; the horizontal axis measures fluorescence from labeled antibody to CD18. The shift of the plot to the left in the patient compared with the control represents a 10-fold decrease in the number of lymphocytes displaying CD18 on their surface.

further compromised because of the vital role of CR3-mediated uptake of these opsonized bacteria by neutrophils. The role of CR4 or CD11c,CD18, the third member of the β_2 integrin family, is less well understood but, like CR3, it binds complement fragments, and it is thought to have a role in uptake of bacteria by macrophages.

The importance of the daily, massive neutrophil emigration into the oral cavity is well illustrated by individuals with leukocyte adhesion deficiency, who invariably develop severe gingivitis when they survive. Another, poorly understood, consequence of the lack of leukocyte emigration is the failure to heal wounds. Delayed separation of the umbilical cord is the earliest manifestation of this defect in wound healing. Subsequently, affected children may develop fistulas (abnormal connecting channels) in their intestine after bacterial infections of the gut.

Bone marrow transplantation has been very successful in rescuing severely affected infants from certain death.

Discussion and questions.

1   *Can you surmise the inheritance pattern of the leukocyte adhesion deficiency from this case?*

It is autosomal recessive. The parents are both healthy; they have had an affected male and an affected female child. These facts lead to the conclusion that the leukocyte adhesion deficiency is inherited as an autosomal recessive trait. In fact the gene encoding CD18 has been mapped to the long arm of chromosome 21 at the position 21q22. An examination of the leukocytes of the parents would reveal decreased expression of CD18 but sufficient for normal function.

2   *In some families with the leukocyte adhesion deficiency, a mild phenotype is observed and leukocytes of affected individuals express 10% of the normal amount of CD18. The mild phenotype is usually due to a splice site defect in the gene encoding CD18, whereas, in the severe form of the disease, there is usually no transcription of the CD18 gene. Do these observations help you to predict the likely effects of treatment with monoclonal antibodies to CD18? Can you suggest situations in which such treatment might be therapeutically useful?*

Monoclonal antibodies to CD18 can induce a mild phenotype of leukocyte adhesion deficiency. Such monoclonal antibodies have been used to prevent graft rejection in recipients of kidney grafts and, when administered before bone marrow transplantation, can prevent graft-versus-host disease.

3   *Can you suggest why the homing of T cells to lymphoid tissue is normal in patients with leukocyte adhesion deficiency and why these individuals do not suffer from susceptibility to opportunistic infections?*

T cells express the β₁ integrin VLA-4 (Fig. 13.6), and its interaction with VCAM-1 seems to be sufficient to enable T cells to home and function normally. VLA-4 is not expressed on neutrophils and macrophages, which are much more dependent on β₂ integrins for their adhesion to other cells. B cells also home normally in leukocyte adhesion deficiency and this is probably due to the integrin α₄β₇, which is not defective in this disease.

4   What manifestation of defective wound healing occurred in this case?

Luisa's brother had delayed separation of the umbilical cord and Luisa developed an infection at the site of umbilical cord separation. The role that neutrophils and macrophages play in wound healing is not well understood. Nevertheless, it is apparent from cases with leukocyte adhesion deficiency that the movement of white blood cells into wounds is vital to normal tissue repair.

5   Another rare form of leukocyte adhesion deficiency has been described in which affected infants cannot convert mannose to fucose. Can you explain why the failure to convert mannose to fucose should cause leukocyte adhesion deficiency?

Fucose is a defining determinant of the sialyl-Lewis x element, which is the ligand whereby white blood cells bind to selectins (see Fig. 13.1). Consequently, infants who cannot make this determinant have very high white blood cell counts because the leukocytes cannot roll on the endothelium to begin the process of leukocyte emigration from the bloodstream.

6   Shortly after Luisa received a bone marrow transplant from her mother it was ascertained that she had complete chimerism of the myeloid and lymphoid cells. How was this done?

After the transplant her leukocytes were found to express CD18.

7   Why was Luisa treated with busulfan, anti-thymocyte serum and cyclophosphamide before the transplantation?

This was done to destroy her abnormal cells and create 'space' for the transplanted cells. You may recall that such preparation was not performed in the case of severe combined immunodeficiency because the lymphoid compartment was already devoid of T cells and thus 'space' was available for the transplanted cells without any treatment.

8   Why were the responses of Luisa's T cells to PHA and Con A depressed?

PHA and Con A are so-called non-specific T-cell mitogens. The mitogenic response requires cell–cell interactions that depend upon the interaction of LFA-1 with ICAM-1, as well as VLA-4 with VCAM-1. In leukocyte adhesion deficiency, one of these interactions is missing.

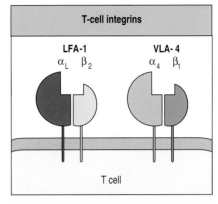

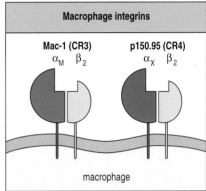

Fig. 13.6 T cells and phagocytes express different integrins. Integrins are heterodimeric proteins containing a β chain, which defines the class of integrin, and an α chain, which defines the different integrins within a class. The α chain is larger than the β chain and contains binding sites for divalent cations that may be important in signaling. LFA-1, a β₂ integrin, and VLA-4, a β₁ integrin, are expressed on T cells and are important in the migration and activation of these cells. Macrophages and neutrophils express all three members of the β₂ integrin family, LFA-1, Mac-1 (also known as CR3), and p150.95 (also known as CR4). Like LFA-1, Mac-1/CR3 binds the immunoglobulin superfamily molecules ICAM-1, ICAM-2, and ICAM-3, but in addition it is a complement receptor (for the fragment iC3b). The function of p150.95/CR4, which also binds complement, is unknown.

CASE 14 Wiskott–Aldrich Syndrome

Role of the actin cytoskeleton in T-cell function.

Many functions of T cells require the directed reorganization of the cell's cytoskeleton, in particular the actin cytoskeleton. The eukaryotic cell cytoskeleton as a whole consists of actin filaments, microtubules and inter-mediate filaments. It provides a framework for the internal structural organization of the cell and is also essential for cell movement, cell division and many other cell functions. In T cells, as in other animal cells, the actin cytoskeleton is found mainly as a meshwork of actin filaments immediately underlying the plasma membrane (Fig. 14.1). The actin cytoskeleton is a dynamic structure and can undergo rapid reorganization by depolymerization and repolymerization of actin filaments. As we shall see in the case of

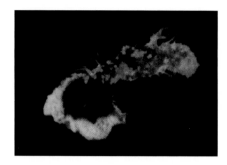

Fig. 14.1 An activated cytotoxic T cell moving over a surface. The actin cytoskeleton is stained green. The red staining indicates the lytic granules containing cytotoxic proteins. Photograph courtesy of Gillian Griffiths.

Wiskott–Aldrich syndrome, an inability of T cells to reorganize their actin cytoskeleton when required has profound effects on their function and thus on immune function as a whole.

Reorganization within the T-cell cortical actin cytoskeleton takes place when T cells interact with B cells or other target cells through cell-surface receptors. The functions of T cells in immune defense all involve interactions with other cells that are initiated by direct cell–cell contact via cell-surface receptors. For example, helper T cells interact with B cells through cell-surface receptors to stimulate B-cell proliferation and the subsequent differentiation into antibody-producing plasma cells. T-cell–B-cell interactions are also involved in isotype switching and generation of memory cells, whereas cytotoxic T-cell killing of virus-infected target cells also involves direct contact with the target cell.

The cytoskeleton is linked to cell-surface receptors in the plasma membrane so that events occurring at the membrane can affect cytoskeleton reorganization. For example, crosslinking of T-cell antigen receptors and co-receptors by antigen:MHC complexes leads to their aggregation at one pole of the T cell, with an accompanying concentration of the actin cytoskeleton at that point also. Binding of a helper T cell to a B cell through its T-cell receptors' binding to antigen:MHC complexes on the B-cell surface also leads to a reorganization of the actin cytoskeleton locally in the zone of contact, which in turn causes a microtubule-dependent mechanism to focus the secretory apparatus of the T cell on the point of contact with the B cell (Fig. 14.2); the release of cytokines from the T cell is thus directed to the contact point. Similar cytoskeletal reorganizations occur when a cytotoxic T cell contacts its target cell.

Fig. 14.2 Binding of a helper T cell to an antigen-binding B cell causes a reorganization of the cytoskeleton in the T cell. Engagement of the T-cell receptor causes the T cell to express the CD40 ligand (CD40L), which binds to CD40 on the surface of the B cell (top panel). The cross-linking of the receptors at the point of contact leads to a reorganization of the cortical actin cytoskeleton, shown here by the redistribution of the protein talin (shown in red in top left and both center panels), which is associated with the actin cytoskeleton. Subsequent reorientation of the secretory apparatus of the T cell leads to cytokine release at the point of contact with the antigen-binding B cell, as shown by staining for IL-4 (bottom right panel), which shows the IL-4 (green) confined to the space between the B cell and the T cell. Photographs courtesy of A Kupfer.

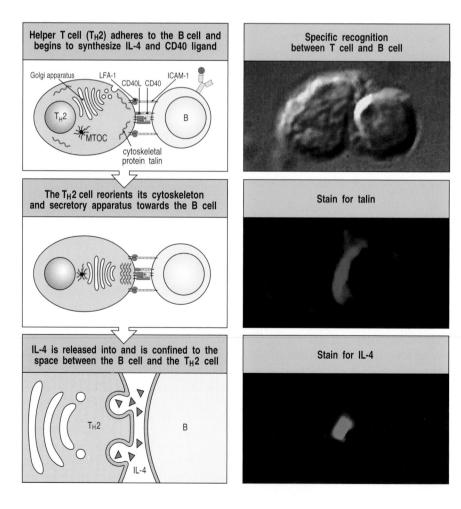

Many other T-cell functions depend on the actin cytoskeleton. Like many other animal cells, T lymphocytes move in a crawling 'ameboid' fashion. The movement of T lymphocytes as they emigrate from the thymus into the blood vessels and subsequently 'home' from the bloodstream into lymphoid tissue requires the active participation of the actin cytoskeleton. Cell division induced by the activation of T cells by antigen or by non-specific mitogens also involves the actin cytoskeleton in that the cell is divided into two by the action of a contractile ring formed of actin filaments and myosin.

T cells from patients with Wiskott–Aldrich syndrome are deficient in all these normal cellular abilities and, in particular, seem unable to interact successfully with B cells and other target cells.

The case of Austin Sloper: the consequences of a failure of T-cell–B-cell interaction.

Austin Sloper was first referred to the Children's Hospital at 2 years of age with a history of recurrent infections, eczema, asthma, and episodes of bloody diarrhea. He had been a full-term baby and appeared quite normal at birth. Routine immunizations of DPT (diphtheria, pertussis, and tetanus) and oral polio vaccine had been given at 3, 4, and 5 months of age without any untoward consequences. At 6 months his mother noticed eczema developing on his arms and legs; this was treated with 1% cortisone ointment. The eczema became infected with *Staphylococcus epidermitidis,* and petechiae (small skin hemorrhages) appeared in the eczematous areas as well as on unaffected areas of skin. By 2 years old he began to have frequent middle ear and respiratory infections, including pneumonia as confirmed by a chest radiograph. Between respiratory infections he started to wheeze and was found to have asthma.

Normally developed male infant; recurrent infections, eczema, asthma

A blood analysis showed normal levels of hemoglobin of 11.5 g dl^{-1} and a normal white blood cell count of 6750 μl^{-1}. The proportions of leukocytes of different types (eg granulocytes and lymphocytes) were also normal. The platelet count was, however, low (thrombocytopenia), at 40,000 μl^{-1} (normal 150,000–350,000), and the platelets were abnormally small (Fig. 14.3). Because of the combination of repeated infections, eczema, and thrombocytopenia with small platelets, the Wiskott–Aldrich syndrome (WAS) was diagnosed.

Further immunological investigations at this time revealed levels of IgG of 750 mg dl^{-1} (normal), IgM 25 mg dl^{-1} (decreased), IgA 475 mg dl^{-1} (increased), and IgE 750 international units ml^{-1} (increased). Austin's red blood cells were type O but no anti-A or anti-B isohemagglutinins were present in his serum. He made no immune response to immunizations with a pneumococcal vaccine (composed of capsular polysaccharides) or to polyribose phosphate (PRP) vaccine (composed of the

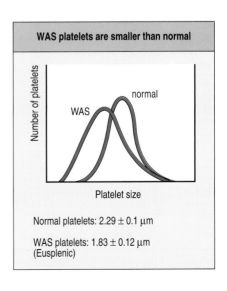

WAS platelets are smaller than normal

Number of platelets

normal

WAS

Platelet size

Normal platelets: 2.29 ± 0.1 μm

WAS platelets: 1.83 ± 0.12 μm (Eusplenic)

Fig. 14.3 Platelet sizing. Blood was drawn from a normal subject and from a patient with WAS into sodium citrate, an anti-coagulant. The blood was then centrifuged and allowed to sediment at 1g and the platelet-rich plasma was removed. This was diluted in sterile buffer and scanned in a particle sizer. The mean diameter of normal platelets is 2.29 ± 0.1 μm. The mean diameter of platelets from patients with the Wiskott–Aldrich syndrome is 1.83 ± 0.12 μm. Data kindly supplied by Dianne Kenney.

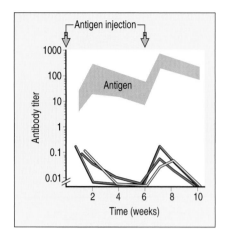

Fig. 14.4 Impaired primary and secondary immune responses in Wiskott–Aldrich syndrome. The figure shows the response of three patients with Wiskott–Aldrich syndrome to the antigen bacteriophage φX174 given intravenously. Their titer of anti-bacterio-phage antibodies was measured at 2, 4, and 6 weeks after the injection. At 6 weeks, a second dose of bacteriophage was administered. The antibody titer was then measured at 1, 2, and 4 weeks after the second injection. The shaded area represents the 66% confidence limits of the response of normal children to the same antigen. Data kindly supplied by Hans Ochs.

Thrombocytopenia with small platelets; Wiskott–Aldrich?

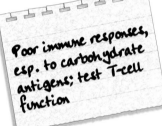

Poor immune responses, esp. to carbohydrate antigens; test T-cell function

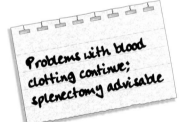

Problems with blood clotting continue; splenectomy advisable

cell-wall carbohydrate of *Haemophilus influenzae*). His titer of antibodies against tetanus toxid (a protein) was 0.1 unit ml⁻¹ (a borderline normal result). He was given two intravenous doses of the bacteriophage φX174 as a test antigen 6 weeks apart. A primary response was elicited by the first injection but the secondary response was poor (Fig. 14.4).

He had normal numbers of circulating B lymphocytes (11% of peripheral blood lymphocytes) and T lymphocytes (85%) with a normal distribution of CD4⁺ and CD8⁺ T cells. However, when his T cells were stimulated with the mitogens concanavalin A and phytohemagglutinin a poor proliferative response was obtained. Nor did his T cells respond to the mitogenic effect of anti-CD3 antibodies.

As he grew older, Austin developed frequent severe nosebleeds. At these times his platelet count was found to have dropped to 10,000 μl⁻¹ and he required platelet transfusions to stop the nosebleeds. He frequently developed purpuric lesions (black and blue spots due to hemorrhaging) on the skin.

At 14 years of age his spleen was removed (splenectomy), after which his platelet count immediately rose to 253,000 μl⁻¹ and has remained in the normal range, although the platelets continue to be of smaller than normal size (see legend to Fig. 14.3). His absolute T-cell count had progressively decreased with age from 3500 μl⁻¹ to 1250 μl⁻¹. He was started on intravenous gamma globulin at a dose of 30 g every 3 weeks (at the time he weighed 60 kg) and was advised to take penicillin before visits to the dentist or during respiratory infections.

Further allergies developed, particularly to nuts. His asthma persisted and he was advised to take theophylline 300 mg twice daily. At age 16 he developed severe chickenpox with systemic symptoms of varicella hepatitis, which was successfully treated with Acyclovir. As he grew older and continued to be treated with gamma globulin the frequency of infections decreased. Austin is now in his his mid-30s, works full time as a teacher and is in reasonably good health.

After Austin, his mother had three more male children: Jonathan, Thornton and Matthew. Jonathan and Matthew both had WAS; Thornton was normal. Mrs Sloper had five brothers, all of whom are normal; she had no sisters and no daughters. At age 32 Jonathan Sloper developed a B-cell lymphoma. Although this was successfully treated with cytotoxic drugs he died of sepsis (bloodstream infection) due to a Gram-negative enterobacillus. Matthew Sloper developed glomerulonephritis due to deposition of immune complexes in his kidney during his 20s and at age 34 received a successful kidney transplant from a cadaver donor. His renal function continues to be normal 3 years after the transplant, and he works full time as a business executive. He has not had a splenectomy and his platelet count remains around 45,000 μl⁻¹. He has no problems with abnormal bleeding.

Wiskott–Aldrich syndrome.

Austin Sloper and his two affected brothers illustrate between them all the principal features of WAS. The syndrome was first described by Wiskott in 1937 in Munich when he observed three male infants with bloody diarrhea, eczema and thrombocytopenia with small platelets. In 1947 Aldrich and his colleagues studied a large Dutch-American family in Minnesota and established the X-linked inheritance of the disease in five generations of affected males. The immunodeficiency that accompanies WAS was not appreciated until 1968, when it was described by Blaese and Waldmann and by Cooper and Good.

The defective gene responsible for WAS was mapped to the short arm of the X chromosome and the normal gene has now been cloned. This encodes a protein named the Wiskott–Aldrich syndrome protein or WASP, which was found to have homology with actin-binding cytoskeletal proteins involved in the reorganization of the actin cytoskeleton in white blood cells and platelets. WASP is expressed only in white blood cells and megakaryocytes (from which platelets are derived), which explains the restriction of its effects to immune system and blood clotting functions. In patients with WAS, T cells and platelets are defective in number and function. T-cell movement, capacity for cell division, capping of antigen receptors, and reorientation of the cytoskeleton on engagement with other cells are all impaired. B cells and phagocytes seem to be normal in function.

Abnormally small platelets are a prerequisite for a diagnosis of WAS. In this disease there is a defect in platelet production from megakaryocytes in the bone marrow and an increased destruction of blood platelets by the spleen. It is not known what the spleen recognizes as abnormal in WAS platelets; in any event the spleen causes microfragmentation of the platelets so that they become much smaller than usual. A lot of platelet 'dust' (tiny membrane-bounded cytoplasmic fragments) is found in the blood of WAS patients. Splenectomy greatly alleviates this problem, as we saw in Austin's case. The spleen also causes microfragmentation of T lymphocytes in WAS. The T cells lose their microvilli and assume a characteristically bald appearance (Fig. 14.5).

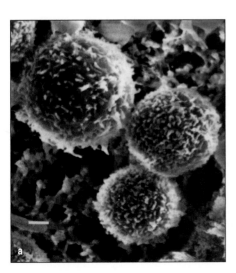 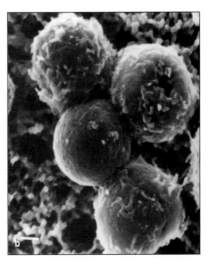

Fig. 14.5 Loss of microvilli on T cells in Wiskott–Aldrich syndrome. Scanning electron micrographs of normal lymphocytes (panel a) and lymphocytes from a patient with Wiskott–Aldrich syndrome (panel b). Note that the normal lymphocyte surface is covered with abundant microvilli, which are sparse or absent from the patient's lymphocytes. Photographs courtesy of Dianne Kenney.

Patients with WAS have increased susceptibility to both pyogenic bacterial infections and opportunistic infections. Among the latter, severe chickenpox (varicella), herpes simplex and molluscum contagiosum (a viral infection of the skin, particularly skin involved with eczema) are frequently encountered. The increased susceptibility to such viral infections may be at least partly due to the impaired cytotoxic function of CD8$^+$ T cells in WAS; the impairment seems to be in their inability to attach to target cells. Antibody formation, particularly to carbohydrate antigens, is defective, as we saw for Austin, who could not respond to pneumococcal polysaccharides and the capsular polysaccharide of *H. influenzae*. He also did not make antibodies to the blood group antigens, which are complex carbohydrates.

The defects in antibody production, including the impaired secondary response, reflect a general failure of T-cell–B-cell interactions. The failure to produce antibodies against capsular polysaccharides and blood group antigens in WAS is of particular interest as it confirms that, in humans, the production of antibodies to complex linear polysaccharides is not independent of T-cell help. In humans these anti-polysaccharide antibodies are normally of the IgG2 subclass; they fix complement and protect against pyogenic infections, which are often due to capsulated bacteria. Normally, the production of IgG2 antibodies would have involved B-cell–T-cell interactions that resulted in an isotype switch.

The importance of anti-polysaccharide antibodies was highlighted by the problems with infection encountered in the case of X-linked agammaglobulinemia (see Case 2), whereas the importance of being able to switch to IgG isotypes is illustrated by the case of Hyper IgM syndrome (see Case 3). The fact that the non-production of polysaccharide antibodies in WAS patients is due to T-cell defects was confirmed by early attempts at bone marrow transplantation to correct WAS. A few patients were converted into partial T-cell chimeras, which was sufficient to clear their eczema and enable them to make antibodies to carbohydrate antigens. Subsequently, complete success at replacing the T-cell population has been achieved by bone marrow transplantation after irradiation to wipe out the abnormal cells in the host.

Although Austin's T-cell number seemed normal when tested at 2 years old, after around 6 years of age the total T-cell number in affected males begins to decline. As this happens, the B cells undergo a general polyclonal expansion. Eventually this B-cell expansion can become monoclonal and B-cell lymphomas appear, as happened in one of Austin's affected brothers. As affected males are now living longer because of improved therapy (WAS used to be fatal in the first decade of life), more are developing B-cell lymphomas in their 30s and 40s. These progressions are reminiscent of what is observed in patients with AIDS.

The role of WASP in cytoskeletal reorganization is not yet understood. WASP has actin-binding domains and also binds the small G protein, Cdc42. It also binds other proteins involved in cytoskeletal reorganization. Mice with 'knock-outs' of the gene encoding WASP have defective homing of lymphocytes and defective receptor capping in T lymphocytes (Fig. 14.6), as well as defective responses of T cells to mitogens and anti-CD3 antibodies. B-cell function seems to be normal. These mice may serve as an animal model for gene therapy for this disease. The high frequency of immune-complex disease and allergy in WAS remains to be explained. So does the persistent finding of low IgM and high IgA and IgE, as though these patients make excessive amounts of interleukin-4. The explanations for these findings are not yet obvious.

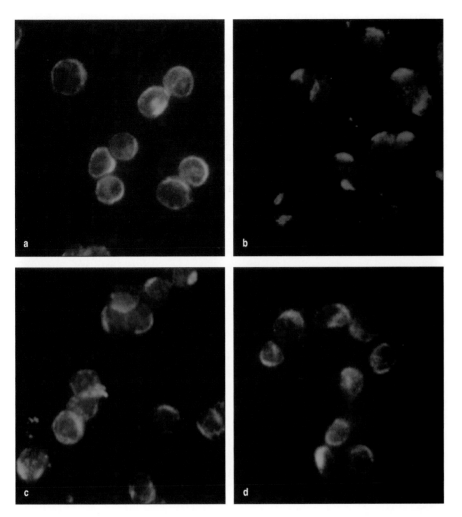

Fig. 14.6 Failure of T-cell capping in WASP knockout mice. T cells were obtained from mice in which the WASP gene had been knocked out by homologous recombination (WASP$^{-/-}$), and from normal mice of the same strain. The T cells were treated with anti-CD3ε antibody, which binds to the T-cell receptor:CD3 complex, and then examined with a fluorescent antibody to CD3. Normal resting T cells (panel a) show capping of the T-cell receptor after anti-CD3ε treatment (panel b). Panel c shows resting T cells from WASP$^{-/-}$ mice, and the failure of cap formation (panel d) after stimulation of the T-cell receptor. Photographs courtesy of Scott Snapper.

Discussion and questions.

1 *All five of Mrs Sloper's brothers were normal. What does this suggest about the source of Mrs Sloper's mutation?*

This strongly suggests that her mother did not have a mutation at the *WAS* locus on her X chromosome, but rather that Mrs Sloper was a new mutant. If her mother had been a carrier, there would have been only a 1 in 32 (2^5) chance of the mother not having an affected child among 5 boys.

2 *When Mrs Sloper and other female heterozygous carriers of WAS are examined for randomness of X-chromosome inactivation, non-random inactivation of the WAS-bearing X chromosome is found in all the blood cell lineages—monocytes, eosinophils, basophils, neutrophils, B lymphocytes and CD4$^+$ and CD8$^+$ T lymphocytes. When hematopoietic stem cells are isolated from these women they also exhibit non-random X inactivation. How might this be explained?*

Signals that direct the maturation of hematopoietic stem cells into the various lineages are transmitted by their contact with stromal cells in the bone marrow. Presumably this interaction, like T-cell–B-cell interaction, requires cytoskeletal reorientation, and thus will be impaired in cells containing an active affected chromosome. The stem cells bearing an active normal X chromosome thus have a survival advantage.

3 *Human infants do not respond to polysaccharide antigens until they are 18 months old. An examination of the B-cell repertoire for receptors specific for polysaccharide antigens reveals that none can be detected until 18 months of age. Patients with WAS also do not respond to polysaccharide antigens. What do you think you would find if you examined their B-cell receptor repertoire?*

Polysaccharide-specific B cells would be present in normal numbers. The failure of response to polysaccharide antigens in WAS results from the failure of T-cell–B-cell collaboration, not from a block in B-cell maturation.

4 *Can you devise a strategy that might induce isotype switching in WAS B cells to overcome the lack of T-cell–B-cell collaboration?*

You might try to give antibody against the B-cell cell-surface protein CD40 along with the immunogen. Ligation of CD40 by the CD40 ligand borne by activated T cells is a signal for a resting B cell to start dividing and to undergo isotype switching. The antibody should act like the CD40 ligand and induce isotype switching in B cells (see Case 3).

5 *Austin Sloper takes theophylline to improve his asthma symptoms. What is theophylline and how does it work?*

Theophylline is one of the methyl xanthines (like caffeine) that inhibits the phosphodiesterase responsible for the dephosphorylation of cyclic AMP. Like the β_2 agonists used in bronchial asthma (see Case 16), it raises the intracellular cyclic AMP level and this induces the relaxation of smooth muscle around the airways.

CASE 15   Chronic Granulomatous Disease

A specific failure of phagocytes to produce H_2O_2 and superoxide.

After microorganisms have been opsonized with antibody and complement and taken up by phagocytic cells, they are killed rapidly within phagocytic vacuoles, or phagosomes (Fig. 15.1). The microbicidal actions of phagocytes occur through a variety of mechanisms (Fig. 15.2). One of the most important of these involves the production of hydrogen peroxide and superoxide radicals, which raises the pH of phagosomes and thereby enables the enzymes that are released into them to attack ingested microbes. A complex enzyme, NADPH oxidase, catalyzes the initial reaction, which results in the generation of superoxide radicals:

Topics bearing on this case:

Microbial action of phagocytes

Chronic inflammation

Leukocyte migration

Fig. 15.1 Activation of a phagocyte.
Microbes (red) are ingested by a phagocyte and enter the cytoplasm in a phagocytic vacuole, or phagosome. When activated by a T_H1 cell, the macrophage becomes activated. The phagosome fuses with a lysosome containing microbicidal enzymes. The enzymes are released into the phagosome, where they kill and degrade the microbe. The elimination of microbes from the vesicles of activated macrophages can be seen in the light micrographs (bottom row) of resting (left) and activated (right) macrophages infected with *M. tuberculosis*. The cells have been stained with an acid-fast red dye to reveal the presence of the microbes, which are prominent in the resting macrophages but have been eliminated from the activated macrophages. Photographs courtesy of G Kaplan.

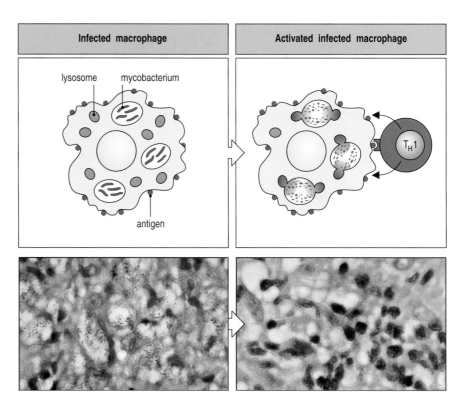

$$NADPH + 2O_2 \longrightarrow NADP^+ + 2O_2^- + H^+$$

The superoxide dismutase then catalyzes the synthesis of hydrogen peroxide:

$$2H^+ + 2O_2^- \longrightarrow H_2O_2 + O_2$$

NADPH oxidase is a large enzyme complex in the phagosome membrane that is assembled from component parts in the membrane and the cytosol in response to a phagocytic stimulus. Not all of the components of this multi-subunit complex have been defined clearly but at least two, $p47^{phox}$ and $p67^{phox}$, reside in the cytoplasm in unstimulated phagocytes, whereas two others together comprise the membrane complex cytochrome b_{558}. The two

Fig. 15.2 Ingestion of bacteria triggers the production or release of many bactericidal agents in phagocytic cells.
Most of these agents are found in both macrophages and neutrophilic polymorpho-nuclear leukocytes (neutrophils). Some of them are cytotoxic; others, such as lacto-ferrin, work by binding essential nutrients and preventing their uptake by bacteria. The same agents can be released by phagocytes interacting with large, antibody-coated surfaces such as parasitic worms or host tissues. As these mediators are also toxic to host cells, phagocyte activation can cause extensive tissue damage in infection.

Class of mechanism	Specific products
Acidification	pH=~3.5 – 4.0, bacteriostatic or bacteriocidal
Toxic oxygen-derived products	Superoxide O_2^-, hydrogen peroxide H_2O_2, singlet oxygen $^1O_2^{\bullet}$, hydroxyl radical $OH^{\bullet}$, hypohalite OCl
Toxic nitrogen oxides	Nitric oxide NO
Antimicrobial peptides	Defensins and cationic proteins
Enzymes	Lysozyme—dissolves cell walls of some Gram-positive bacteria. Acid hydrolases—further digest bacteria
Competitors	Lactoferrin (binds Fe) and vitamin B12 binding protein

Fig. 15.3 The NADH oxidase complex.
NADH is a large, multisubunit enzyme complex whose assembly in the phagocyte membrane is triggered by a phagocytic stimulus. For simplicity, only four major components are shown here: gp91^{phox} and p21^{phox}, which together form the membrane complex known as cytochrome b$_{558}$, and p67^{phox} and p47^{phox}, which normally reside in the cytoplasm. On delivery of a phagocytic stimulus to the cell, p47^{phox} becomes hyperphosphorylated and binds to p67^{phox} and other cytosolic components, and these components migrate to the membrane to form the complete complex with cytochrome b$_{558}$.

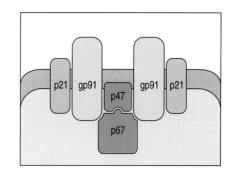

chains of cytochrome b$_{558}$ are a heavy chain, gp91^{phox}, and a light chain, p21^{phox}, which contains the catalytic site (Fig. 15.3). The genes encoding p47^{phox}, p67^{phox}, and p21^{phox} map to autosomal chromosomes, whereas gp91^{phox} is encoded on the short arm of the X chromosome. Four different genetic defects affect various components of the NADPH oxidase enzyme but they all result in a disease with a common phenotype, called chronic granulomatous disease (CGD). The commonest form of the disease is X-linked CGD, which is caused by mutations in the gene encoding gp91^{phox}.

The diagnosis of CGD can be ascertained easily by taking advantage of the metabolic defect in the phagocytic cells. A dye, called nitro blue tetrazolium (NBT), is pale yellow and transparent. When it is reduced, it becomes insoluble and turns a deep purple colour. To test for CGD, a drop of blood is suspended on a slide and is provided with a phagocytic stimulus (for example, phorbol myristate acetate (PMA), an activator of the oxidase); a drop of NBT is added to the blood. Neutrophils, which are the main phagocytes in blood, take up the NBT and the PMA at the same time. In normal blood the NBT is reduced to a dark purple, insoluble formazan, easily seen in the phagocytic cells; in CGD blood no dye reduction is seen (Fig. 15.4).

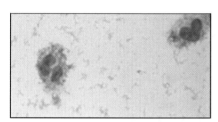

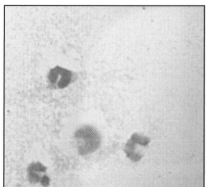

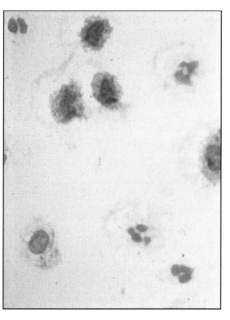

Fig. 15.4 Neutrophils stimulated with phorbol myristate acetate in the presence of nitro blue tetrazolium (NBT) dye.
The left upper panel shows two normal neutrophils stained purple. The left lower panel shows neutrophils from a patient with CGD that failed to reduce NBT. The neutrophils in the right panel are from a heterozygous carrier of chronic granulomatous disease. Half the neutrophils have reduced NBT (purple) and half do not stain. Photographs courtesy of P Newberger.

The case of Randy Johnson: a near death from exotic bacteria and fungi.

Fifteen-year-old boy with impending respiratory failure.

Repeated respiratory infections.

Randy Johnson, a 15-year-old high-school student, had a summer job working with a gardening crew. His job entailed spreading bark mulch in garden beds. At the end of August of that summer he rapidly developed severe shortness of breath, a persistent cough and chest pain. Because of impending respiratory failure he was admitted to the hospital. A radiological examination of Randy's chest was performed and this revealed the presence of large 'cotton ball' densities in both lungs (Fig. 15.5). One of these lesions was aspirated with a fine needle and when the aspirate was stained numerous aspergillus hyphae were seen. A culture of the aspirate grew *Aspergillus fumigatus*.

Randy was started on intravenous amphotericin B and assisted mechanical ventilation through a tracheotomy. He slowly improved over the course of 2 months. During this time in the hospital, he contracted two further respiratory infections with *Pseudomonas aeruginosa* and *Streptococcus faecalis*. These infections were also treated with appropriate antibiotics.

At the time of his initial admission to the hospital, Randy's white blood cell count was 11,500 μl^{-1} (normal 5000–10,000 μl^{-1}). He had 65% neutrophils, 30% lymphocytes, and 5% monocytes: these proportions are normal.

Because of his infections, which were unusual for a seemingly healthy 15-year-old adolescent, his serum immunoglobulins were measured. The serum IgG level was 1650 mg dl^{-1} (upper limit of normal 1500 mg dl^{-1}). IgM and IgA were in the high normal range at 250 and 175 mg dl^{-1} respectively. Because no defect could be found in his humoral immunity, his white cell function was tested with an NBT slide test. His white blood cells failed to reduce NBT.

Randy had four older sisters and two brothers. They all had an NBT test performed on their blood. One, his brother Ralph, aged 9 years, also failed to reduce NBT. His mother reported that Ralph had had a perirectal abscess in infancy but had otherwise been well.

Randy's mother and one sister had a mixed population of neutrophils in the NBT test; about half the phagocytes reduced NBT and the other half did not (see Fig. 15.4). His other three sisters and other brother gave normal NBT reduction.

Further studies with Randy's and Ralph's granulocytes revealed that the rate of hydrogen peroxide production was 1.61 and 0.28 nmol min^{-1} 10^{-6} cells (normal 6.35 nmol min^{-1} 10^{-6} cells). The cytochrome b content in both brothers' granulocytes was <1.0 pmol mg^{-1} protein (normal control 101 pmol mg^{-1} protein). Both Ralph and Randy were started on treatment with injections of interferon-γ.

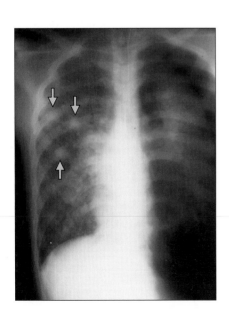

Fig. 15.5 A radiograph of aspergillus pneumonia. Arrows point to 'cotton ball' densities characteristic of fungi.

Chronic granulomatous disease (CGD).

Randy and Ralph Johnson present a mild phenotype of X-linked chronic granulomatous disease. Males affected with this defect usually show undue susceptibility to infection in the first year of life. The most common infections are pneumonia, infection of the lymph nodes (lymphadenitis), and abscesses of the skin and of the viscera such as the liver. Granuloma formation occurs

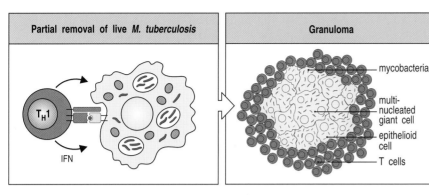

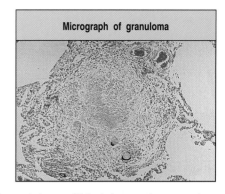

Fig. 15.6 Granulomas form when an intracellular pathogen or its constituents cannot be eliminated completely. The pathogen illustrated here is the intracellular bacterium *Mycobacterium tuberculosis*. In the top panel, mycobacteria (red) are shown resisting the effects of macrophage activation by a T_H1 CD4 T cell. The middle panel is a schematic diagram of a granuloma, which forms as the result of a localized inflammatory response that develops because of the sustained activation of chronically infected phagocytes. The granuloma consists of a central core of infected macrophages, and may include multinucleated giant cells, which are fused macrophages, surrounded by large macrophages often called epithelioid cells. These may, in turn, be surrounded by T cells, many of which are CD4-positive. Granulomas frequently form in CGD because macrophages are unable to destroy the bacteria they ingest and infected macrophages accumulate. The right panel shows a micrograph of a granuloma. Photograph courtesy of J Orrell.

when microorganisms are opsonized and ingested by phagocytic cells but the infection cannot be cleared. The persistent presentation of microbial antigens by phagocytes induces a sustained cell-mediated response by CD4 T cells, which recruit other inflammatory cells and set up a chronic local inflammation called a granuloma. This can occur in normal individuals in response to intracellular bacteria that colonize macrophages (Fig. 15.6); in CGD, where the intracellular microbicidal mechanisms are defective, many infections can persist and set up chronic inflammatory reactions resulting in granulomas. The fungi and bacteria that are most frequently responsible for infections and granuloma formation in CGD are listed in Fig. 15.7.

It was discovered quite by chance that interferon-γ improves the resistance of males with X-linked CGD to infection. The basis for this effect is still not understood. This cytokine does not increase superoxide or peroxide production in the deficient phagocytes; its positive effects remain to be determined.

Discussion and questions.

1 *50% of Randy's mother's neutrophils have reduced the NBT and 50% have not (see Fig. 15.4). How do you explain this?*

There has been random inactivation of the X chromosomes in Randy's mother's neutrophils. Therefore 50% of her neutrophils have a normal X chromosome and 50% have an X chromosome bearing the CGD defect. That half does not reduce NBT, whereas the half bearing the normal X chromosome does.

2 *Children with CGD do not have problems with pneumococcal infections. Could you hypothesize why this is so?*

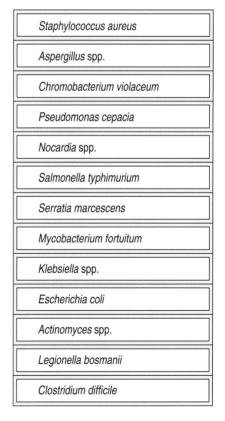

| *Staphylococcus aureus* |
| *Aspergillus* spp. |
| *Chromobacterium violaceum* |
| *Pseudomonas cepacia* |
| *Nocardia* spp. |
| *Salmonella typhimurium* |
| *Serratia marcescens* |
| *Mycobacterium fortuitum* |
| *Klebsiella* spp. |
| *Escherichia coli* |
| *Actinomyces* spp. |
| *Legionella bosmanii* |
| *Clostridium difficile* |

Fig. 15.7 Table of fungal and bacterial infectious agents most commonly responsible for infections in CGD.

Streptococcus pneumoniae, the pneumococcus, does not produce catalase and is thus far less resistant to intracellular killing than microorganisms that are catalase producers.

3 *The immunoglobulin levels in Randy's blood were somewhat elevated. How do you explain this?*

Because of persistent antigenic stimulation he is making more immunoglobulins (antibodies) than a normal person. In fact all chronic infections, such as malaria, result in hypergammaglobulinemia.

4 *How does chronic granulomatous disease differ from leukocyte adhesion deficiency?*

In leukocyte adhesion deficiency, there is a defect in the mobility of the leukocytes so that their emigration from the circulation to sites of infection is impeded. The microbicidal capacity of phagocytes in leukocyte adhesion deficiency is unimpaired. Children with leukocyte adhesion deficiencies therefore never develop inflammatory lesions, which are caused by activated phagocytes, and are equally susceptible to pneumococcal and other bacterial infections (see Question 2). In contrast, the leukocytes in chronic granulomatous disease have normal mobility and can reach sites of infection in a normal manner but cannot efficiently destroy bacterial infections when they reach them. Hence the development of abscesses and granulomas.

CASE 16　Allergic Asthma

Adaptive immunity underlying allergic disease.

As we saw with contact sensitivity to poison ivy in Case 8, adaptive immune responses can be elicited by antigens that are not associated with infectious agents. Inappropriate immune responses to otherwise innocuous foreign antigens result in allergic or hypersensitivity reactions. These unwanted responses can be serious.

Allergic reactions occur when an already sensitized individual is re-exposed to the same innocuous foreign substance or allergen. The first exposure generates allergen-specific antibodies and/or T cells; re-exposure to the same allergen, usually by the same route, leads to an allergic reaction (Fig. 16.1). Once an individual is sensitized, the allergic reaction becomes worse with each subsequent exposure, which not only produces allergic symptoms but also increases the level of antibody and T cells.

The allergic response that we shall discuss in this case is an example of a type I hypersensitivity reaction. Type I reactions involve the activation of helper CD4 T_H2 cells, IgE antibody formation, mast cell sensitization and recruitment of eosinophils. The IgE antibodies formed in sensitized individuals bind to and occupy high-affinity Fcε receptors (FcεRI) on the surface of tissue mast cells and basophils (Fig. 16.2). When the antigen is encountered again, it crosslinks these bound IgE molecules, which triggers the immediate release of mast cell granule contents, particularly histamine and various enzymes that increase blood flow and vascular permeability. This is the early phase of an immediate allergic reaction.

Within 12 hours of contact with antigen, a late-phase reaction occurs (Fig. 16.3). Arachidonic acid metabolism in the mast cell generates prostaglandins and leukotrienes, which further increase blood flow and vascular permeability

Topics bearing on this case:
IgE-mediated hypersensitivity
Differential activation T_H1 and T_H2 cells
Inflammatory reactions
Radioimmunoassay
Skin tests for hypersensitivity
Tests for immune function

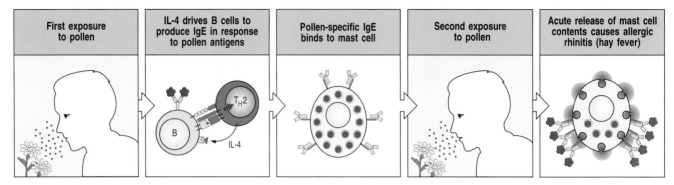

Fig. 16.1 Allergic reactions require prior exposure to the allergen. In this example, the first exposure to pollen induces the production of IgE anti-pollen antibodies, driven by the production of IL-4 by helper T cells (T$_H$2). The IgE binds to mast cells via FcεRI. Once enough IgE antibody is present on mast cells, exposure to the same pollen induces mast-cell activation and an acute allergic reaction, here allergic rhinitis or hay fever. Allergic reactions require pre-sensitization to the antigen or allergen, and several exposures may be needed before the allergic reaction is initiated.

(Fig. 16.4). Cytokines such as IL-3, IL-4, IL-5, and tumor necrosis factor-α (TNF-α) are also produced by both activated mast cells and helper T cells and these further prolong the allergic reaction. The mediators and cytokines released by mast cells and helper T cells cause an influx of monocytes, more T cells, and eosinophils into the site of allergen entry. The late-phase reaction is dominated by this cellular infiltrate. The cells of the infiltrate, particularly the eosinophils, make a variety of products that are thought to be responsible

Fig. 16.2 Crosslinking of IgE antibody on mast cell surfaces leads to rapid release of inflammatory mediators by the mast cells. Mast cells are large cells found in connective tissue that can be distinguished by secretory granules containing many inflammatory mediators. They bind stably to monomeric IgE antibodies through the very high-affinity Fcε receptor (FcεRI). Antigen crosslinking of the bound IgE antibody molecules triggers rapid degranulation, releasing inflammatory mediators into the surrounding tissue. These mediators trigger local inflammation, which recruits cells and proteins required for host defense to sites of infection. It is also the basis of the acute allergic reaction causing asthma, hay fever, and the life-threatening response known as systemic anaphylaxis (see Case 24). Photographs courtesy of A M Dvorak.

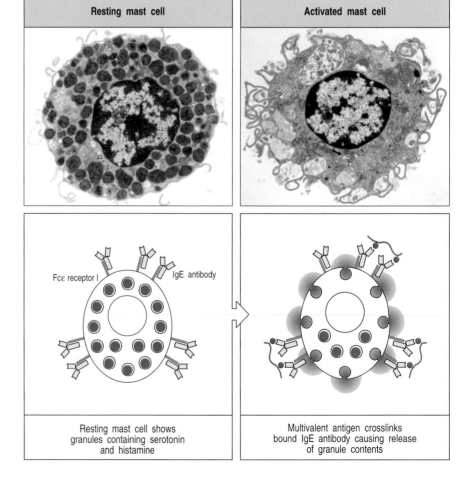

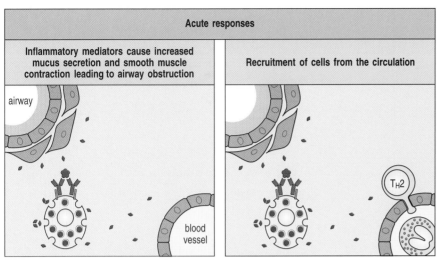

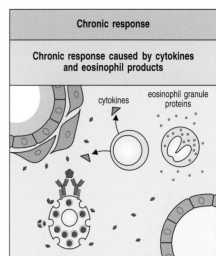

Fig. 16.3 The acute response in allergic asthma leads to T$_H$2-mediated chronic inflammation of the airways. In sensitized individuals, crosslinking of specific IgE on the surface of mast cells by inhaled allergen triggers them to secrete inflammatory mediators, causing bronchial smooth muscle contraction and an influx of inflammatory cells, including eosinophils, and T$_H$2 lymphocytes. Activated mast cells and T$_H$2 cells secrete cytokines that also augment eosinophil activation and degranulation, which causes further tissue injury and influx of inflammatory cells. The end result is chronic inflammation, which might then cause irreversible damage to the airways.

for much of the tissue damage and overproduction of mucus that is associated with chronic allergic reactions. Mast cells are also reactivated and thus the inflammatory reaction is perpetuated.

Approximately 15% of the population suffer from IgE-mediated allergic diseases. Many common allergies are caused by inhaled particles containing foreign proteins (or allergens) and result in allergic rhinitis (eg hay fever), asthma, and allergic conjunctivitis (Fig. 16.5). In asthma, the allergic inflammatory response increases the hypersensitivity of the airway not only to re-exposure to allergen but also to non-specific agents. So exercise, pollutants, and cold air can cause asthma in people without allergies (this is known as intrinsic asthma; allergic asthma is known as extrinsic asthma).

Mast-cell mediators	
Preformed inflammatory mediators in secretory granules	Histamine, proteoglycans Serine proteases: chymases, tryptases Carboxypeptidases
Lipid-derived mediators (newly formed)	Leukotriene B$_4$, leukotrienes C$_4$, D$_4$ and E$_4$ (SRS-A), prostaglandin D$_2$, platelet activating factor
Mast cell-derived cytokines	Proinflammatory cytokines: TNF-α, IL-1α, IL-1β, IL-6 MIP family
Mitogenic cytokines and/or growth factors	IL-3, IL-4, IL-5, IL-10, GM-CSF
Immunomodulatory cytokines	IL-1α and IL-1β, IL-4, IL-10, IFN-γ

Fig. 16.4 Mediators released by mast cells upon crosslinking surface IgE.

Fig. 16.5 IgE-mediated reactions to extrinsic antigens. All IgE-mediated responses involve mast-cell degranulation, but the symptoms experienced by the patient can be very different depending on whether the allergen is injected, inhaled, or eaten, and depending on the dose of the allergen.

IgE-mediated allergic reactions			
Syndrome	Common allergens	Route of entry	Response
Systemic anaphylaxis	Drugs Serum Venoms Peanuts	Intravenous (either directly or following rapid absorption)	Edema Increased vascular permeability Tracheal occlusion Circulatory collapse Death
Acute urticaria (wheal-and-flare)	Insect bites Allergy testing	Subcutaneous	Local increase in blood flow and vascular permeability
Allergic rhinitis (hay fever)	Pollens (ragweed, timothy, birch) Dust-mite feces	Inhaled	Edema of nasal mucosa Irritation of nasal mucosa
Asthma	Pollens Dust-mite feces	Inhaled	Bronchial constriction Increased mucus production Airway inflammation
Food allergy	Shellfish Milk Eggs Fish Wheat	Oral	Vomiting Diarrhea Pruritis (itching) Urticaria (hives) Anaphylaxis (rarely)

The case of Frank Morgan: a 14-year-old boy with chronic asthma and rhinitis.

14-year-old boy with persisting wheezing. History of chronic asthma and rhinitis.

Frank Morgan was referred by his pediatrician to the allergy clinic at age 14 years because of persistent wheezing for 2 weeks. His symptoms had not responded to frequent inhalation treatment (every 2–3 hours) with a bronchodilator, the β₂-adrenergic agent albuterol (salbutamol).

This was not the first time that Frank had had respiratory problems. His first attack of wheezing occurred when he was 3 years old, after a visit to his grandparents who had recently acquired a dog. He had similar attacks of varying severity on subsequent visits to his grandparents. Beginning at age 4, he had attacks of coughing and wheezing every spring (April and May) and towards the end of the summer (second half of August and September). A sweat test at age 5 to rule out cystic fibrosis, a possible cause of chronic respiratory problems, was within the normal range.

As Frank got older, gym classes, basketball and soccer games, and just going out of doors during the cold winter months could bring on coughing and sometimes wheezing. He had been able to avoid wheezing induced by exercise by inhaling albuterol 15–30 minutes before taking exercise. Frank had frequently suffered from a night cough and his colds had often been complicated by wheezing.

Frank's chest symptoms have been treated as needed with bronchodilators such as oral theophylline and/or oral or inhaled albuterol. During the past 10 years Frank has been admitted three times to hospital for treatment of his asthma with inhaled bronchodilators and intravenous steroids. He has also made many emergency visits with severe asthma attacks. He has had maxillary sinusitis at least three times, ascertained by the presence of fluid in the cavity of his maxillary sinuses, which shows up on the radiographs, each episode associated with exacerbation of his asthma and a green nasal discharge.

Since he was 4 years old Frank has also suffered from intermittent sneezing, nasal itching and nasal congestion (rhinitis), which always worsens upon exposure to cats and dogs and in the spring and late summer. His face also swells up on exposure to a cat or a dog. The nasal symptoms have been treated as needed with oral antihistamines with moderate success. Frank had eczema as a baby but this cleared up by age 5.

Family history revealed that Frank's only sibling (a 10-year-old sister), his mother and his maternal grandfather all have asthma. Frank's mother, father, and paternal grandfather suffer from allergic rhinitis.

When he arrived at the allergy clinic, Frank was thin and unable to breathe easily. He had no fever. The nasal mucosa was severely congested, and wheezing could be heard over all the lung fields. Lung function tests were consistent with obstructive lung disease with a reduced peak flow rate (PFR) of 180 liter min^{-1} (normal >350–400 liter min^{-1}) and expiratory volume in the first second of expiration (FEV$_1$) was reduced to 50% of that predicted for his sex, age, and height. A chest radiograph showed hyperinflation of the lungs and increased markings around the airways (Fig. 16.6).

A complete blood count was normal except for a high number of circulating eosinophils (1200 μl^{-1}; normal range <400 μl^{-1}). Serum IgE was high at 1750 international units ml^{-1} (normal less than 200 units ml^{-1}). Radioallergsorbent assays (RAST) for antigen-specific IgE revealed IgE antibodies to dog and cat dander, dust mites, tree grass, and ragweed pollens in Frank's serum. Levels of immunoglobulins IgG, IgA, and IgM were normal. Histological examination of Frank's nasal fluid showed the presence of eosinophils.

Frank was given immediate inhalation treatment of 0.25 ml of the bronchodilator Ventolin (albuterol sulfate) at the clinic, after which he felt better, his PFR rose to 400 liter min^{-1} and his FEV$_1$ rose to 65% of predicted. He was sent home on a 1-week course of the corticosteroid prednisone, taken orally in two doses daily, totalling 1 mg kg^{-1} body weight per day. He was told to inhale albuterol three times a day with extra doses if needed but not more than every 4 hours. He was also put on inhaled Intal (disodium cromoglycate), a drug that decreases mast cell granule release and inhibits allergen-induced airway reactivity. To relieve his nasal congestion, Frank was given disodium cromoglycate and the steroid beclomethasone (Beconase, Vancenase) to inhale through the nose, and was advised to use an oral antihistamine as needed. He was asked to return to the clinic 2 weeks later for follow-up, and for immediate hypersensitivity skin tests to try to detect to which antigens he was allergic (Fig. 16.7).

On the next visit Frank had no symptoms except for a continually stuffy nose. His PFR and FEV$_1$ were normal. He was maintained on inhaled disodium cromoglycate and beclomethasone, and advised to inhale albuterol as needed. Skin tests for type I hypersensitivity were positive for multiple tree and grass pollens, dust mites, and dog and cat dander. He was advised to avoid contact with cats and dogs. To reduce his exposure to dust mites the pillows and mattresses in his room were covered with zippered plastic covers. Rugs, stuffed toys, and books were removed from his bedroom. He was also started on immunotherapy with injections of grass, tree, and ragweed pollens, and house dust mite antigens, to try to reduce his sensitivity to these antigens.

A year and a half later Frank's asthma continues to be stable with occasional use of albuterol during upper respiratory tract infections and in the spring. His rhinitis and nasal congestion now require much less medication.

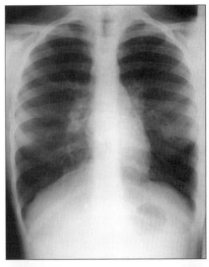

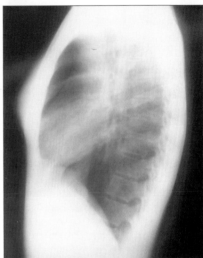

Fig. 16.6 Chest radiographs of a patient with asthma. Top: anteroposterior (A-P) view. Bottom: lateral view. The volume occupied by the lungs spans eight to nine rib spaces instead of the usual seven in the A-P view and indicates hyperinflation. The lateral view shows an increased A-P dimension also reflecting hyperinflation. The bronchial markings are accentuated and can be seen to extend beyond one-third of the lung fields. This indicates inflammation of the airways.

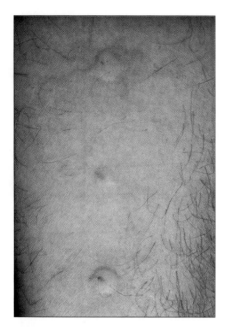

Fig. 16.7 An intradermal skin test.
The photograph was taken 20 min after intradermal injections had been made with ragweed antigen (top), saline (middle) and histamine (bottom). A central wheal (raised central swelling), reflecting increased vascular permeability, surrounded by a flare (red area), reflecting increased blood flow, is observed at the sites where the ragweed antigen and the histamine-positive control were introduced. The small wheal at the site of saline injection is due to the volume of fluid injected into the dermis.

Allergic asthma.

Millions of adults and children like Frank suffer from allergic asthma. About 70% of patients with asthma have a family history of allergy. This genetic predisposition to develop allergic diseases is called atopy. Wheezing and coughing are the main symptoms of asthma, and both are due to the forced expiration of air through airways that have become temporarily narrowed by constriction of smooth muscle as a result of the allergic reaction. As a consequence of the narrowed airways, air gets trapped in the lung and the lung volume is increased during an attack of asthma.

Once asthma is established, an asthma attack can be triggered not only by the allergen but by viral infection, cold air, exercise, or pollutants such as sulfur dioxide. This is due to a general hyperirritability of the airway, which leads it to constrict in response to non-specific shocks and irritants, reducing the air flow. The degree of hyperirritability can be measured by determining the threshold dose of inhaled metacholine (a cholinergic agent) or histamine that results in a 20% reduction in airway flow. Airway irritability correlates positively with serum IgE levels.

Although asthma is a reversible disease, a severe attack can be fatal. The mortality from asthma has been rising alarmingly in recent years. Three classes of drugs are commonly used to treat it. Disodium cromoglycate reduces airway irritability by inhibiting the release of chemical mediators (eg histamine) from mast cells and therefore inhibits both the immediate and late phases of the allergic reaction. The precise mechanism of action of this drug is unknown. β_2 agonists (eg albuterol) bind to the β_2-adrenergic receptor, which is expressed on the surface of bronchial smooth muscle cells. β_2 agonists relax smooth muscle and thus rapidly relieve airway constriction, and are helpful in treating the immediate phase of the allergic reaction in the lungs. Anti-inflammatory corticosteroids (eg oral prednisone and inhaled beclomethasone) inhibit the cells involved in airway inflammation and are most useful in the late phase of the allergic reaction. The treatment of allergic asthma also includes minimizing exposure to allergens, and trying to desensitize the patient by immunotherapy.

Discussion and questions.

1   *Explain the basis of Frank's chest tightness and radiograph findings.*

During inspiration, the negative pressure on the airways causes their diameter to increase, allowing inflow of air. During expiration, the positive expiratory pressure tends to narrow the airways. This narrowing is exaggerated when the airway is inflamed and bronchial smooth muscle is constricted, as in asthma. This causes air to be trapped in the lungs with an increase in residual lung volume at the end of expiration. Breathing at high residual lung volume means more work for the muscles and increased expenditure of energy; this results in the sensation of tightness in the chest. The high residual lung volume is also the cause of the hyperinflated chest observed on the chest radiograph. The peribronchial inflammation in asthma causes bronchial marking around the airways.

2 Explain the failure of Frank's asthma to improve despite the frequent use of bronchodilators, and his response to steroid therapy.

Chronic allergic asthma is not simply due to constriction of the smooth muscles that surround the airway. It is largely due to the inflammatory reaction in the airway, which consists of cellular infiltration, increased secretion of mucus, and swelling of the bronchial tissues. This explains the failure of bronchodilators, which dilate smooth muscles, to maintain an open airway and their failure to reverse completely the decreased air flow during Frank's acute attacks. Steroids are therefore given to combat the inflammatory reaction of the late-phase response.

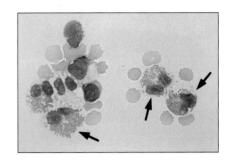

Fig. 16.8 Eosinophilia. The arrows indicate eosinophils.

3 Many members of Frank's family are predisposed to develop IgE-mediated allergic responses, a condition known as atopy. What is the basis for this familial predisposition ?

Analysis of the genetic linkage of atopy to various polymorphic DNA markers suggests that this trait maps to chromosome 5q11 in the area containing the genes for IL-4, IL-5, and IL-9, cytokines that are involved in regulating the IgE response. Because IL-4 is required to induce B cells to undergo isotype switching to IgE, it is possible that the underlying cause of the high IgE levels in atopic individuals is the excessive production of IL-4.

4 Eosinophilia (Fig. 16.8) is often detected in the blood and in the nasal and bronchial secretions of patients with allergic rhinitis and asthma. What is the basis for this finding?

Allergic individuals have a tendency to respond to allergens with an immune response skewed to producing T_H2 cells rather than T_H1. The cells produce the interleukins IL-4 and IL-13, cytokines that induce IgE production in humans. T_H2 cells also make IL-5, which is essential for eosinophil maturation. Furthermore, activated T cells and bronchial epithelial cells secrete eotoxin, which attracts eosinophils in the airways. Production of IL-4 and IL-5 by T_H2 cells responding to allergens in atopic individuals explains the frequent association of IgE antibody response and eosinophilia in these patients.

5 What is the basis of the wheal-and-flare reaction that appeared 20 minutes after Frank had had a skin test for hypersensitivity to ragweed pollen?

IgE-mediated hypersensitivity to an allergen is tested for by injecting a small amount of the allergen intradermally. In allergic individuals, this is followed within 10–20 minutes by a wheal-and-flare reaction at the site of injection (see Fig. 16.7), which subsides within an hour. The wheal-and-flare reaction is due mainly to the release of histamine by mast cells in the skin. This increases the permeability of blood vessels and leakage of their contents into the tissues, resulting in the swollen wheal; dilation of the fine blood vessels around the area produces the diffuse red 'flare' seen around the wheal. This reaction is almost completely inhibited by antagonists to the histamine type 1 receptor, the major histamine receptor expressed in the skin.

6 Frank called 24 hours after his skin test to report that redness and swelling had recurred at several of the skin test sites. Explain this observation.

The recurrence of the redness and swelling at the site of previous immediate allergic reactions represents the late-phase response characterized by a cellular infiltrate.

7   *Frank asks to be skin tested for sensitivity to rabbits because he wishes to have a rabbit as a pet. What do you do?*

Because Frank has made IgE antibodies to numerous allergens, it is very likely that should he become exposed to a rabbit, he would mount an IgE antibody response to rabbit allergens, which would probably cause allergic symptoms. A negative skin test would probably simply reflect the fact that Frank has not yet been exposed to rabbit allergens. Therefore Frank should be advised not to get a rabbit regardless of the result of the skin test.

8   *How would the immunotherapy that Frank received help to alleviate his allergies?*

Repeated administration of relatively high doses of allergen by subcutaneous injection is thought to favor antigen presentation by antigen-presenting cells that produce IL-12. This results in the induction of inflammatory CD4 T cells (T_H1 cells) rather than T_H2 cells. The presence of T_H1 cells tends to lead to an IgG antibody response rather than an IgE response, as the T_H1 cells produce interferon-γ, which prevents further isotype switching to IgE. The IgG antibody competes for antigen with IgE. Furthermore, IgG bound to allergen inhibits mast cell activation (via FcϵRI) and B-cell activation (via surface Ig) by allergen because of inhibitory signals delivered subsequent to binding of Fcγ receptors on these cells. This is thought to be one mechanism damping down the allergic response. Another is no further boosting of IgE production as IL-4 and IL-13 are not secreted. Existing IgE levels themselves may not fall by much, as IFN-γ does not affect B cells that have already switched to IgE production.

9   *Although atopic children are repeatedly immunized with protein antigens such as tetanus toxoid, they almost never develop allergic reactions to these antigens. Explain.*

Most human allergy is caused by a limited number of inhaled protein allergens that elicit a T_H2 response in genetically predisposed individuals. These allergens are relatively small, highly soluble protein molecules that are presented to the immune system by the mucosal route at very low doses. It has been estimated that the maximum exposure to ragweed pollen allergens is less than 1 µg per year. It seems that transmucosal presentation of very low doses of allergens favors activation of IL-4-producing T_H2 cells and is particularly efficient at inducing IgE responses. The dominant antigen-presenting cell type in the respiratory mucosa expresses high levels of co-stimulatory B7.2 molecules. Expression of B7.2 on antigen-presenting cells is thought to favor the development of T_H2 cells. In contrast, injection of antigen subcutaneously in large doses, as occurs on vaccination, results in antigen uptake in the local lymph nodes by a variety of antigen-presenting cells and favors the development of T_H1 cells, which inhibit antibody switching to IgE.

CASE 17 | Myasthenia Gravis

The immune response turns against the host.

The specific adaptive immune response can, in rare instances, be mounted against self antigens and cause autoimmune disease. Injury to body tissues can result from antibodies directed against cell-surface or extracellular-matrix molecules, from antibodies bound to molecules circulating in the plasma that deposit as immune complexes, or from clones of T cells that are reactive with self antigens. A special class of autoimmune diseases are caused by autoantibodies to cell-surface receptors (Fig. 17.1). Graves' disease and myasthenia gravis are two well-studied examples. Graves' disease is caused by autoantibodies to the receptor on thyroid cells for thyroid-stimulating hormone (TSH), secreted by the pituitary gland. In this disease, the binding of the autoantibody to the TSH receptor acts like TSH itself, and stimulates the thyroid gland to produce thyroid hormones. In myasthenia gravis, the opposite effect is observed: antibodies to the acetylcholine receptor at the neuro-muscular junction impede the binding of acetylcholine and stimulate internalization of the receptor, thereby blocking the transmission of nerve impulses by acetylcholine (Fig. 17.2).

Topics bearing on this case:

Humoral autoimmunity

Transfer of maternal antibodies

Mechanisms for breaking tolerance

Fig. 17.1 Autoimmune diseases caused by antibody to surface or matrix antigens. These are known as Type II autoimmune diseases. Damage by IgE-mediated responses (Type I) does not occur in autoimmune disease. In most Type II diseases, autoantibodies bind to the cell surface or extracellular matrix and target them for destruction by phagocytes (often with the help of complement) and/or natural killer cells. A special class of autoimmune diseases is caused by autoantibodies that bind cellular receptors and either stimulate or block their normal function. Immune-complex disease (Type III) is discussed in Case 23. T-cell mediated disease (Type IV) is discussed in Cases 8 and 21.

Some common Type II autoimmune diseases caused by antibody to surface or matrix antigens		
Syndrome	**Autoantigen**	**Consequence**
Autoimmune hemolytic anemia (see Case 20)	Rh blood group antigens, I antigen	Destruction of red blood cells by complement and phagocytes, anemia
Autoimmune thrombocytopenia purpura	Platelet integrin gpIIb:IIIa	Abnormal bleeding
Goodpasture's syndrome	Non-collagenous domain of basement membrane collagen type IV	Glomerulonephritis Pulmonary hemorrhage
Pemphigus vulgaris (see Case 22)	Epidermal cadherin	Blistering of skin
Acute rheumatic fever	Streptococcal cell wall antigens, antibodies cross-react with cardiac muscle	Arthritis, myocarditis, late scarring of heart valves
Graves' disease	Thyroid-stimulating hormone receptor	Hyperthyroidism
Myasthenia gravis	Acetylcholine receptor	Progressive weakness
Insulin-resistant diabetes	Insulin receptor (antagonist)	Hyperglycemia, ketoacidosis
Hypoglycemia	Insulin receptor (agonist)	Hypoglycemia

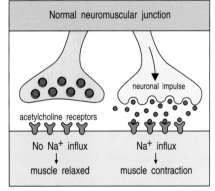

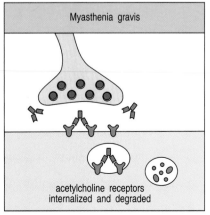

Myasthenia gravis means severe (gravis) muscle (my) weakness (asthenia). It was first realized that this disease was an autoimmune disease when an immunologist immunized rabbits with purified acetylcholine receptors in order to obtain antibodies to this receptor. He noticed that the rabbits developed floppy ears, like the droopy eyelids (ptosis) that are the most characteristic symptom of myasthenia gravis in humans. It was then shown that patients with this disease indeed have antibodies to the acetycholine receptor. It has been known for a long time that pregnant women with myasthenia gravis transfer the disease to their newborn infants. As IgG is the only maternal serum protein that crosses the placenta from mother to fetus, the occurrence of neonatal myasthenia gravis is clear evidence that myasthenia gravis is caused by an IgG antibody.

Fig. 17.2 Autoantibodies to the acetylcholine receptor weaken the reception of the signal from nerve ends that cause the muscle cell to contract. At the neuromuscular junction, acetylcholine is released from stimulated neurons and binds to acetylcholine receptors, triggering muscle contraction. The acetylcholine is destroyed rapidly by the enzyme acetylcholinesterase after release. In myasthenia gravis, auto-antibodies to the acetylcholine receptor induce its endocytosis and degradation, and prevent muscles from responding to neuronal impulses.

The case of Mr Weld: from floppy ears to droopy eyelids.

Mr Weld is a 71-year-old retired engineer. He developed double vision (diplopia), which gradually worsened over the course of 4 months and caused him to seek medical attention. He had been in good health and active his entire life.

On examination, the doctor noticed that Mr Weld had ptosis of both eyelids so that they covered the upper third of the irises of his eyes. When the doctor asked Mr Weld to look to the right and then to the left, he noticed limitations in the ocular movements of both eyes, as shown in Fig. 17.3.

The remainder of the neurological examination was normal. No other muscle weakness was found during the examination.

A radiological examination of the chest was performed, and it was normal. There was no evidence in the radiograph of enlargement of the thymus gland. A blood sample was taken from Mr Weld and his serum was tested for antibodies to the acetylcholine receptor. The serum contained 6.8 units of antibody to the acetylcholine receptor (normal <0.5 units).

Mr Weld was told to take pyridostigmine, an inhibitor of cholinesterase.

His double vision steadily improved but he developed diarrhea from the pyridostigmine, and this limited the amount he could take.

Ptosis of both eyelids. Limitation in ocular movements of both eyes.

Prescribe Pyridostigmine.

Gaze to right	Straight ahead	Gaze to left
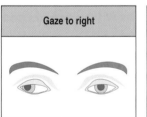	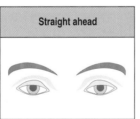	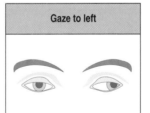

Fig. 17.3 Diagram of ocular movement limitation.

Three years later

Mr Weld developed a severe respiratory infection. Soon after, his ptosis became so severe that he had to lift his eyelids by taping them with adhesive tape.

His diplopia recurred and his speech became indistinct. He developed difficulty with chewing and swallowing food. He could only tolerate a diet of soft food and it would take him several hours to finish a meal.

The neurologist noted that he now had weakness of his facial muscles and his tongue, and the abnormality in ocular movements again became apparent.

Because of the diarrhea he was only able to tolerate a quarter of the prescribed dose of pyridostigmine. He also developed difficulty in breathing. His vital capacity (the amount of air he could exhale in one deep breath) was low, at 3.5 liters.

Mr Weld was admitted to the hospital and treated with azathioprine. Thereafter he showed steady improvement.

His ptosis and diplopia improved remarkably and he was able to eat normally. His vital capacity returned to normal and was measured to be 5.1 liters.

Symptoms recur during severe respiratory infection, difficulty chewing and swallowing vital capacity 3.5 l.

Myasthenia gravis.

Mr Weld illustrates the features of a common type of myasthenia gravis, which is observed in older people and is called the oculobulbar form because it principally involves the muscles of the eye. In younger people, the disease is much more severe and the antibody titers to the acetylcholine receptor are higher. In very severe cases, impaired breathing may even cause death, and difficulty in swallowing can cause aspiration of food particles into the lung. Chest radiographs of younger people with myasthenia gravis frequently reveal enlargement of the thymus gland, and sometimes tumors of the thymus (thymoma). Removal of the thymus gland (thymectomy) in younger patients may lead to rapid improvement in their symptoms. The relationship of the thymus gland to myasthenia gravis is a mystery, and it is equally unclear why thymectomy improves the symptoms of this disease. Abnormalities of the thymus are not observed in older patients and thymectomy does not help them.

Discussion and questions.

1 *Newborn infants of mothers with myasthenia gravis exhibit symptoms of myasthenia gravis at birth. How long would the disease be likely to last in these infants?*

It would be likely to last 1–2 weeks. The infant has the disease because maternal IgG antibodies to the acetylcholine receptor have crossed the placenta from the maternal circulation to the fetal circulation. The infant is not synthesizing these autoantibodies; he or she has acquired the disease passively by transfer of the antibodies (Fig. 17.4). The maternal IgG antibodies bind to the acetylcholine receptors in the baby, and the complex of the receptor with bound IgG antibodies is internalized into the cell and degraded. Within 10–15 days all the maternal IgG antibodies to the acetylcholine receptor are adsorbed from the babies' blood and the symptoms abate.

2 *Pyridostigmine is an ideal drug for the treatment of myasthenia gravis. It inhibits the enzyme cholinesterase, which normally cleaves and inactivates acetylcholine. In this way, pyridostigmine prolongs the*

Fig. 17.4 Antibody-mediated autoimmune disease can appear in the infants of affected mothers as a consequence of transplacental transfer of immunoglobulin G.

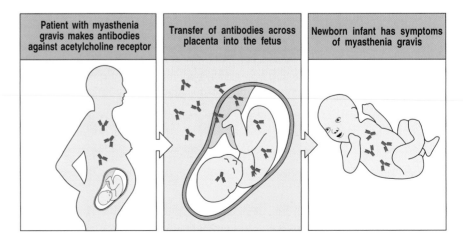

| Patient with myasthenia gravis makes antibodies against acetylcholine receptor | Transfer of antibodies across placenta into the fetus | Newborn infant has symptoms of myasthenia gravis |

biological half-life of acetylcholine. Unfortunately, it also causes diarrhea by raising the amount of acetylcholine in the intestine. Acetylcholine binds to the muscarinic receptors in the intestine and increases intestinal motility. Since he could not tolerate full therapeutic doses of pyridostigmine and was getting worse, Mr Weld was given azathioprine and showed marked improvement. What did the azathioprine do? What would concern you about prolonged use of this drug?

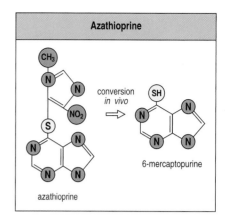

Fig. 17.5 The structure of azathioprine and its active product.

Azathioprine (Fig. 17.5) is an immunosuppressive agent. It is converted in the liver to 6-mercaptopurine, which inhibits DNA synthesis. Thus the growth of rapidly dividing cells, such as B cells and T cells, is inhibited and the immune response is suppressed. Unfortunately the effects of azathioprine are not specific. It suppresses not only the formation of antibodies to the acetylcholine receptor but also all other immune responses. Patients taking azathioprine become susceptible to infections. If used for very prolonged periods it is associated with the development of lymphomas. The reasons for this are not well understood.

$\boxed{3}$ *Mr Weld had a severe relapse in his disease following a respiratory infection. Many autoimmune diseases appear to be triggered by infection, and relapses in autoimmune diseases frequently follow an infection. Can you explain how this might happen?*

Fig. 17.6 summarizes ways in which infectious diseases can break self tolerance and induce, or worsen, an autoimmune disease. An infectious agent may expose hidden antigens, or may increase expression of MHC molecules and co-stimulators on tissue cells so as to induce an autoimmune response. B cells already primed to make an autoantibody may receive help from nearby T cells activated by an infection, especially if pathogens become attached to self molecules. Pathogens may induce responses that cross-react with self molecules. Bacterial and viral superantigens can overcome clonal anergy and break tolerance to self antigens. At present, relatively little is known about the induction of human autoimmune disease, and there are only a few examples in which the evidence for any one of these mechanisms is strong.

Mechanism	Disruption of cell or tissue barrier	Infection of antigen-presenting cell	Binding of pathogen to self protein	Molecular mimicry	Superantigen
Effect	Release of sequestered self antigen; activation of non-tolerized cells	Induction of co-stimulatory activity on antigen-presenting cells	Pathogen acts as carrier to allow anti-self response	Production of cross-reactive antibodies or T cells	Polyclonal activation of autoreactive T cells
Example	Sympathetic ophthalmia	Effect of adjuvants in induction of EAE	? Interstitial nephritis	Rheumatic fever ? Diabetes ? Multiple sclerosis	? Rheumatoid arthritis

Fig. 17.6 Ways in which infectious agents can break self tolerance. Because some antigens are sequestered from the circulation, either behind a tissue barrier or within the cell, it is possible that an infection that breaks cell and tissue barriers may expose hidden antigens (first column). A second possibility is that the local inflammation in response to an infectious agent may trigger the expression of MHC molecules and co-stimulators on tissue cells, inducing an autoimmune response (second column). In some cases, infectious agents may bind to self proteins. Because the infectious agent induces a helper T-cell response, any B cell that recognizes the self protein will also receive help. Such responses should be self-limiting once the infectious agent is eliminated, because at this point, the T-cell help will no longer be provided (third column). Infectious agents may induce either T-cell or B-cell responses that can cross-react with self antigens. This is termed molecular mimicry (fourth column). T-cell polyclonal activation by a bacterial superantigen could overcome clonal anergy, allowing an autoimmune process to begin (fifth column). There is little evidence for most of these mechanisms in human autoimmune disease. EAE, experimental autoimmune encephalomyelitis.

CASE 18 | MHC Class II Deficiency

An inherited failure of gene regulation.

The class II molecules of the major histocompatibility complex (MHC) are involved in presenting antigens to CD4 T cells. The peptide antigens that they present are derived from extracellular pathogens and proteins taken up into intracellular vesicles, or from pathogens such as *Mycobacterium* that persist intracellularly inside vesicles. MHC class II molecules are expressed constitutively on antigen-presenting cells, including B lymphocytes, macrophages, and dendritic cells. In humans, together with the MHC class I molecules (see Case 4) they are known as the HLA antigens. They are also expressed on the epithelial cells of the thymus and their expression can be induced on other cells, principally by the cytokine interferon-γ. T cells also express MHC class II molecules when they are activated.

Fig. 18.1 Structure of an MHC class II molecule. Panel a shows a computer graphic representation of the MHC class II molecule HLA-DR1. Panel b is a schematic representation of the molecule. It is composed of two transmembrane glyco-protein chains, α and β, each folded into two protein domains. The antigenic peptide binds in a cleft between the two chains. Photograph courtesy of C Thorpe.

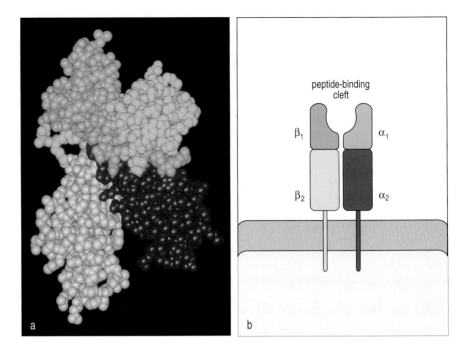

MHC class II molecules are heterodimers consisting of an α chain and a β chain (Fig. 18.1). The genes encoding both chains are located in the MHC on the short arm of chromosome 6 in humans (Fig. 18.2). The principal MHC class II molecules are designated DP, DQ, and DR, and like the MHC class I molecules they are highly polymorphic. Peptides bound to MHC class II molecules can be recognized only by the T-cell receptors of CD4 T cells and not by those of CD8 T cells (Fig. 18.3). MHC class II molecules expressed in the thymus also have a vital role in the intrathymic maturation of CD4 T cells.

Expression of the genes encoding the α and β chains of MHC class II molecules must be coordinated strictly and is under complex regulatory control. The regulation of MHC class II gene expression is not fully understood as it involves the action of transcription factors that are defined only in part. The existence of these transcription factors and a means of identifying them were first suggested by the study of patients with MHC class II deficiency.

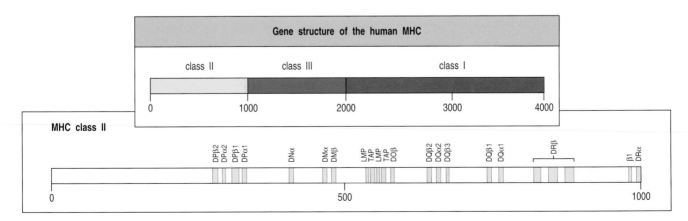

Fig. 18.2 Detailed map of the MHC class II region. The genes for the α and β chains of the HLA-DP, HLA-DR, and HLA-DQ molecules are shown as DPα, DPβ, etc. The situation is complicated as there are two DPα genes and several DPα and DPβ genes, whose expression has to be coordinated.

The case of Helen Burns: a 6-month-old child with a mild form of severe combined immunodeficiency.

Helen Burns was the second child born to her parents. She thrived until 6 months of age when she developed pneumonia in both lungs, accompanied by a severe cough and fever. Blood and sputum cultures for bacteria were negative but a tracheal aspirate revealed the presence of abundant *Pneumocystis carinii*. She was treated successfully with the anti-*Pneumocystis* drug pentamidine and seemed to recover fully.

As her pneumonia was caused by the opportunistic pathogen *Pneumocystis carinii*, Helen was suspected to have severe combined immunodeficiency. A blood sample was taken and her peripheral blood mononuclear cells were stimulated with phytohemagglutinin (PHA) to test for T-cell function by ^{3}H-thymidine incorporation into DNA. A normal T-cell proliferative response was obtained, with her T cells incorporating 114,050 counts min^{-1} of ^{3}H-thymidine (normal control 75,000 counts min^{-1}). Helen had received routine immunizations with orally administered polio vaccine and DPT (diphtheria, pertussis, and tetanus) vaccine at 2 months old. However, in further tests, her T cells failed to respond to tetanus toxin *in vitro*, although they responded normally in the ^{3}H-thymidine incorporation assay when stimulated with allogeneic B cells (6730 counts min^{-1} incorporated compared with 783 counts min^{-1} for unstimulated cells).

When it was found that Helen's T cells could not respond to a specific antigenic stimulus, her serum immunoglobulins were measured and found to be very low. IgG levels were 96 mg dl^{-1} (normal 600–1400 mg dl^{-1}), IgA was 6 mg dl^{-1} (normal 60–380 mg dl^{-1}), and IgM 30 mg dl^{-1} (normal 40–345 mg dl^{-1}).

Helen's white blood cell count was elevated at 20 000 cells µl^{-1} (normal range 4000–7000 µl^{-1}). Of these, 82% were neutrophils, 10% lymphocytes, 6% monocytes, and 2% eosinophils. The calculated number of 2000 lymphocytes µl^{-1} is low for her age (normal >3000 µl^{-1}). Of her lymphocytes, 7% were B cells as determined by an antibody to CD20 (normal 10–12%) and 57% reacted with antibody to the T-cell marker CD3. Of these T cells, 34% were positive for CD8, and 20% were positive for

6-month-old girl with pneumonia. SCID? Do lymphocyte function tests.

Low Ig levels, deficiency of CD4 T cells.

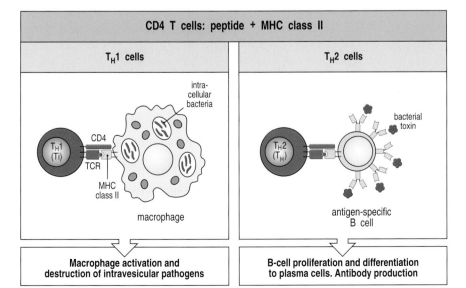

Fig. 18.3 Effector CD4 cells recognize antigens bound to MHC class II molecules. CD4 T cells carry the co-receptor molecule CD4, which binds to MHC class II molecules on the antigen-presenting cell and helps to stabilize T-cell receptor:antigen binding. Effector CD4 T cells fall into two main types: T$_H$1 and T$_H$2. T$_H$1 cells are involved mainly in responding to antigens presented by macrophages, whereas T$_H$2 cells respond to antigen presented by B cells, stimulating the differentiation of B cells to plasma cells and the production of antibodies.

Fig. 18.4 Detection of MHC class II molecules by fluorescent antibody. Helen's transformed B-cell line was examined by using a fluorescent antibody to HLA-DQ and -DR. Helen (left panels) expressed approximately 1% of the amount of MHC class II molecules compared with a transformed B-cell line from a normal control (right panels).

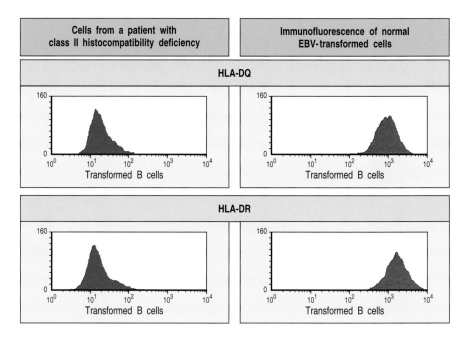

CD4. Thus, at 388 cells μl^{-1} her number of CD8 T cells was within the normal range, but the number of CD4 T cells (228 cells μl^{-1}) was much lower than normal (her CD4 T-cell count would be expected to be twice her CD8 T-cell count). The presence of substantial numbers of T cells, and thus a normal response to PHA, ruled out a diagnosis of severe combined immunodeficiency (see Case 5).

Helen's pediatrician referred her to the Children's Hospital for consideration for a bone marrow transplant, despite the lack of a diagnosis. When an attempt was made to HLA type Helen, her parents and her healthy 4-year-old brother, a DR type could not be obtained from Helen's white blood cells. A long-term culture of her B cells was made by transforming them with Epstein–Barr virus and the transformed B cells were then examined for expression of MHC class I and class II molecules with fluorescent-tagged antibodies. It was found that her B cells did not express HLA-DQ or HLA-DR molecules and a diagnosis of MHC class II deficiency was established (Fig. 18.4).

As her brother did not have the same HLA type as Helen, it was decided to use her mother as a bone marrow donor. Helen was given 1 mg kg^{-1} body weight of cytotoxic drug busulfan every 6 hours for 4 days to depress bone marrow function and then 50 mg kg^{-1} of cyclophosphamide to ablate her bone marrow. The maternal bone marrow was depleted of T cells to diminish the chance of graft-versus-host disease developing and was administered to Helen by transfusion. The graft was successful and immune function was restored.

No HLA-DR type available. Do FACS analysis.

MHC class II deficiency. Bone marrow transplant advisable.

MHC class II deficiency.

MHC class II deficiency is inherited as an autosomal recessive trait. Health problems show up early in infancy. Affected babies present the physician with a mild form of severe combined immunodeficiency (SCID) as they have increased susceptibility to pyogenic and opportunistic infections. However, they differ from SCID patients (see Case 5) in that they have T cells, which can respond to non-specific T-cell mitogens such as PHA and to allogeneic stimuli. Also unlike SCID patients, they do not sustain graft-versus-host disease when given HLA-mismatched blood transfusions. This is because the host tissue has no MHC class II molecules against which the T cells in the graft can react. SCID patients, in contrast, carry MHC II molecules on some of their tissues, against which T cells in the graft will react to cause graft-versus-host disease. They have no T cells at all and therefore cannot reject the graft. Unlike in some other types of immunodeficiency, progressive infection with the attenuated live vaccine strain BCG has not been observed in MHC class II-deficient patients after BCG vaccination against tuberculosis (most cases of MHC class II deficiency have been observed in North African migrants in Europe, where BCG vaccination is routine). This is because mycobacterial antigens derived from BCG can be presented on MHC class I molecules and infected cells can be destroyed by cyotoxic T cells.

Patients with MHC class II deficiency are deficient in CD4 T cells, in contrast to MHC class I deficiency, in which CD8 T cell numbers are very low and the levels of CD4 cells are normal (see Case 4). They also have moderate to severe hypogammaglobulinemia.

Genetic linkage analysis in large extended families with MHC class II deficiency has shown that this condition is not linked to the MHC locus on the short arm of chromosome 6 and that the genes encoding the MHC class II molecules at this locus are normal. Interferon-γ induces the expression of MHC class II molecules on antigen-presenting cells from normal people but fails to induce their expression on the antigen-presenting cells of patients with MHC class II deficiency. This suggested that the defect might lie in the regulation of expression of the MHC class II genes.

The search for the cause of the defect was complicated further by the discovery that MHC class II deficiency in different patients seems to have different causes. B-cell lines isolated from class II-deficient patients do not express MHC class II molecules. However, when B cells from two different patients are fused, MHC class II expression is often observed. The fusion of the two cell lines has corrected the defect. This means that one cell must be able to replace whatever is lacking in the other, and thus the two cells must carry different genetic defects causing the MHC class II deficiency. Pairwise fusions were performed on a large number of cell lines from different patients, and at least four complementation groups were found (Fig. 18.5).

These experiments provided clues that led eventually to the identification of the defect. The lack of MHC class II molecules turns out to result from defects in the transcription factors required to regulate their co-ordinated expression. Three of these transcription factors, which bind to the 5′ regulatory region of the MHC class II genes, have been identified.

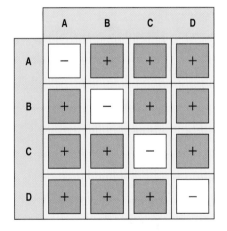

Fig. 18.5 Complementation groups of MHC class II deficiency. B-cell lines isolated from different patients were fused in all pairwise combinations to determine whether they could correct each other's defect. If two cell lines do not correct each other (−), they are in the same complementation group and have the same genetic defect. However, if the defect is corrected (+), the two cell lines belong to two different complementation groups and have two different defects. Four complementation groups, A, B, C, and D, were discovered by this technique.

Discussion and questions.

1   *Why did Helen lack CD4 T cells in her blood?*

The maturation of CD4 T cells in the thymus depends on the interaction of thymocytes with MHC class II molecules on thymic epithelial cells. When the MHC class II genes are deleted genetically in mice, the mice also exhibit a deficiency of CD4 T lymphocytes.

2   *Why did Helen have a low level of immunoglobulins in her blood?*

The polyclonal expansion of B lymphocytes and their maturation to immunoglobulin-secreting plasma cells requires helper cytokines from CD4 T cells, such as interleukin-4. Helen's hypogammaglobulinemia is thus a consequence of her deficiency of CD4 T lymphocytes.

3   *In SCID, lymphocytes fail to respond to mitogenic stimuli. Although Helen was first thought to have SCID, this diagnosis was eliminated by her normal response to PHA and an allogeneic stimulus. How do you explain these findings?*

Helen's T cells, although decreased in number, are normal and are not affected by the defect. They are capable of normal responses to non-specific mitogens and to an allogeneic stimulus in which the antigen is presented by the MHC molecules on the surface of the (non-defective) allogeneic cells and thus does not require to be processed and presented by the defective cells. However, the failure of her lymphocytes to respond to tetanus toxin *in vitro* resulted from the fact that, in this situation, there were no cells that could present antigen on MHC class II molecules to the CD4 T cells.

4   *If a skin graft were to be placed on Helen's forearm do you think she would reject the graft?*

Yes. Helen's T cells would be capable of recognizing the foreign MHC molecules on the grafted skin cells and would reject the graft.

CASE 19 Hemolytic Disease of the Newborn

The adaptive immune response distinguishes different individuals within a species.

Topics bearing on this case:

Antibodies can be produced against almost any substance

ABO blood groups

Rhesus blood group antigens

Hemagglutination assays

Anti-immunoglobulin antibodies are a useful tool for detecting bound antibodies

Antibody can suppress naive B-cell activation

Adaptive immune responses evolved to protect vertebrate species against the world of microorganisms. However, anything discerned as 'non-self' may become a target of such responses, which can also be directed at molecular differences between individuals within a species. An antibody directed against an antigenic determinant present in some members of a species but not others is said to show polymorphic variation, and antibodies directed against such determinants are called alloantibodies. Perhaps the best known alloantibodies are those that are used to determine our blood groups. Individuals who have red cells of type A have alloantibodies that react with the red blood cells of individuals who are type B and vice versa. Individuals who have red blood cells of type AB have no alloantibodies and those with type O red blood cells have antibodies to both A and B red blood cells (Fig. 19.1). These alloantibodies arise from the fact that the capsules of Gram-negative bacteria, which inhabit our gut, bear antigens that stimulate antibodies that cross-react with the carbohydrate antigens of the ABO blood groups.

Fig. 19.1 Hemagglutination is used to type blood groups and match compatible donors and recipients for blood transfusion. Common gut bacteria bear antigens that are similar or identical to blood group antigens, and these stimulate the formation of antibodies to these antigens in individuals who do not bear the corresponding antigen on their own red blood cells (left column); thus, type O individuals, who lack A and B, have both anti-A and anti-B antibodies, whereas type AB individuals have neither. The pattern of agglutination of the red blood cells of a transfusion donor or recipient with anti-A and anti-B antibodies reveals the individual's ABO blood group. Before transfusion, the serum of the recipient is also tested for antibodies that agglutinate the red blood cells of the donor, and vice versa, a procedure called a cross-match, which may detect potentially harmful antibodies to other blood groups that are not part of the ABO system.

Serum from individuals of type	Red blood cells from individuals of type			
	O R–GlcNAc–Gal Fuc	A R–GlcNAc–Gal–GalNAc Fuc	B R–GlcNAc–Gal–Gal Fuc	AB R–GlcNAc–Gal–GalNAc Fuc + R–GlcNAc–Gal–Gal Fuc
O Anti-A and anti-B antibodies	no agglutination	agglutination	agglutination	agglutination
A Anti-B antibodies	no agglutination	no agglutination	agglutination	agglutination
B Anti-A antibodies	no agglutination	agglutination	no agglutination	agglutination
AB No antibodies to A or B	no agglutination	no agglutination	no agglutination	no agglutination

A serious problem is posed frequently by alloantibodies induced by a fetus in the pregnant mother. Alloimmunization most often results from Rhesus (Rh) incompatibility between mother and fetus. Approximately 13% of women are Rh-negative; that is, their red blood cells do not bear the Rh antigen. A woman who is Rh-negative has an 87% chance of marrying an Rh-positive man and their chances of having an Rh-positive baby are very high. Not infrequently, during delivery of the newborn, some blood escapes from the fetal circulation into the maternal circulation and the mother develops alloantibodies to the Rh antigen as a result. During a subsequent pregnancy with an Rh-positive fetus, the maternal IgG alloantibodies cross the placenta and cause hemolysis of the fetal red cells. As we shall see, the consequences of this can be very grave and result in fetal death or severe damage to the newborn infant.

The Rh antigenic determinants are spaced very far apart on the red cell surface. As a consequence, IgG antibodies to the Rh antigen do not fix complement and therefore do not hemolyze red blood cells *in vitro*. For reasons that are less well understood, IgG antibodies to the Rh antigen do not agglutinate Rh-positive red blood cells. Because of this it was very difficult to detect Rh antibodies until Professor Robin Coombs at the University of Cambridge devised a solution to the problem by developing antibodies against human immunoglobulin. He showed that Rh-positive red blood cells coated with IgG anti-Rh antibodies could be taken from a fetus and agglutinated by antibodies to IgG. Furthermore, he showed that when the serum of an alloimmunized woman was incubated with Rh-positive red blood cells, these red blood cells could then be agglutinated by antibody to IgG (Fig. 19.2). The former is called the direct Coombs test and the latter the indirect Coombs test. This application of immunology to a vexing clinical problem led ultimately to a treatment and prevention of the problem.

Fig. 19.2 The Coombs direct and indirect anti-globulin tests for antibody to red blood cell antigens. An Rh-negative (Rh⁻) mother of an Rh-positive (Rh⁺) fetus can become immunized to fetal red blood cells that enter the maternal circulation at the time of delivery. In a subsequent pregnancy with an Rh⁺ fetus, IgG anti-Rh antibodies can cross the placenta and damage the fetal red blood cells. In contrast to anti-Rh antibodies, anti-ABO antibodies are of the IgM isotype and cannot cross the placenta, and so do not cause harm. Anti-Rh antibodies do not agglutinate red blood cells, but their presence can be shown by washing the fetal red blood cells and then adding antibody to human immunoglobulin, which agglutinates the antibody-coated cells. The washing removes unrelated immunoglobulins that would otherwise react with the anti-human immunoglobulin antibody. Anti-Rh antibodies can be detected in the mother's serum in an indirect Coombs test; the serum is incubated with Rh⁺ red blood cells, and once the antibody binds, the red cells are treated as in the direct Coombs test.

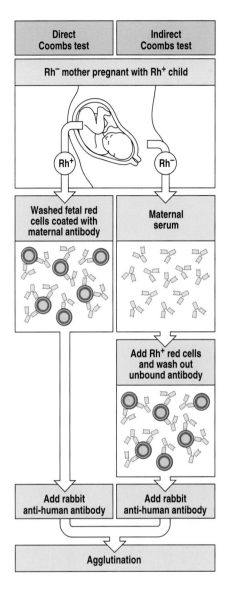

The case of Cynthia Waymarsh: a fetus in immunological distress.

Mrs Waymarsh was 31 years old when she became pregnant for the third time. She was known to have blood group A, Rh-negative red cells. Her husband was also type A but Rh-positive. Their first-born child, a male, was healthy. During her second pregnancy Mrs Waymarsh was noted to have an indirect Coombs titer at a 1:16 dilution of her serum. The fetus was followed closely and the delivery of a healthy baby girl was induced at 36 weeks of gestation.

It was 5 years later when Mrs Waymarsh became pregnant again. At 14 weeks gestation her indirect Coombs titer was 1:8 and at 18 weeks 1:16. Amniotic fluid was obtained at 22, 24, 27, and 29 weeks of gestation and was found to have increasing amounts of bilirubin (a pigment derived from the breakdown of heme, indicating that the fetus' red blood cells were being hemolyzed). At 29 weeks of gestation a blood sample was obtained from the umbilical vein and found to have a hematocrit of 6.2% (normal 45%). (The hematocrit is the proportion of blood that is composed of red cells, and because the volume of white cells is comparatively negligible, this is simply ascertained by centrifuging whole unclotted blood in a tube.) Upon finding that the fetus was profoundly anemic, 85 ml of type O, Rh-negative packed red blood cells were transfused into the umbilical vein. At 30.5 weeks of gestation another sample of blood from the umbilical vein was obtained; the hematocrit was 16.3%. The fetus was transfused with 75 ml of type O, Rh-negative packed red blood cells.

The fetus was examined at weekly intervals for the appearance of hydrops (see below) and none was observed. At 33.5 weeks of gestation the hematocrit of a blood sample from the umbilical vein was 21%, so 80 ml of O Rh-negative packed red blood cells were again transfused into the umbilical vein. At 34.5 weeks of gestation it was determined that the fetus was sufficiently mature to sustain extrauterine life without difficulty; labor was induced and a normal female infant was born. The hematocrit in the umbilical vein blood was 29%. The baby did well and no further therapeutic measures were undertaken.

Hemolytic disease of the newborn.

Although hemolytic disease of the newborn is most commonly the result of alloimmunization with Rh antigen, other red blood cell alloantigens, such as Lewis, Kell, Duffy, Kidd, Lutheran, and still others, may cause alloimmunization. In any case, the maternal IgG antibodies cross the placenta in increasing amounts during the second trimester of pregnancy, and hemolyze the fetal red blood cells. The resulting anemia may become so severe that, if untreated, the fetus goes into heart failure and develops massive edema; this is called hydrops fetalis and results in fetal death. The risk of fetal development of hydrops rises from 10% when the indirect Coombs titer of the mother is 1:16 to 75% when the maternal titer is 1:128. If the anemia is not so severe as to cause hydrops, the affected infant at birth is still massively hemolyzing red blood cells. Now the newborn must dispose of the heme breakdown pigments rapidly because an excessive accumulation of bilirubin results in the deposition of this pigment in the brain, and severe neurological impairments. In response to the profound anemia, the number of red blood cell precursors (erythroblasts) in the spleen, liver and bone marrow expands rapidly; for this reason, hemolytic disease of the newborn has also been called erythroblastosis fetalis.

As we have seen in this case, the extent of hemolysis can be determined easily by obtaining amniotic fluid into which the fetus commences to urinate by 20 weeks of gestation. The quantity of bilirubin excreted into the amniotic fluid correlates with the amount of hemolysis in the fetus. Secondly, the fetus can be followed by ultrasonography for the development of hydrops. Thirdly, the degree of anemia can be ascertained directly, but with some difficulty, by obtaining a sample of blood from the umbilical vein.

It has become possible in the past few decades to eliminate to a very great extent hemolytic disease of the newborn. All Rh-negative women are given 300 μg of purified immunoglobulin specific for the rhesus antigen (Rhogam) at 28 weeks of gestation and again within 72 hours of delivery. The amount of Rhogam in one vial (300 μg) is sufficient to neutralize 30 ml of fetal blood in the maternal circulation. This procedure fails to prevent alloimmunization in only 0.1% of Rh-negative women, such as Mrs Waymarsh. It must be presumed in such failures that the fetus bled more than 30 ml of blood into its mother.

Discussion and questions.

1 | It was stated that the Rh antigens are so sparsely scattered on the red cell surface that IgG molecules bound to the Rh antigens are too separated to fix C1q. Therefore complement-mediated hemolysis cannot be invoked to explain hemolytic disease of the newborn. By what mechanism are the red cells destroyed?

The red blood cells, coated with Rh antibody, adhere tightly to the Fc receptors of macrophages in the red pulp of the spleen, Kupffer cells of the liver, and elsewhere. It turns out that most IgG antibody to the Rh antigen is of the IgG3 or IgG1 subclass, the IgG subclasses that bind most tightly to the high-affinity Fcγ receptor (CD64) (Fig. 19.3). The macrophages destroy the antibody-coated red cells, as they would antibody-coated bacteria (Fig. 19.4).

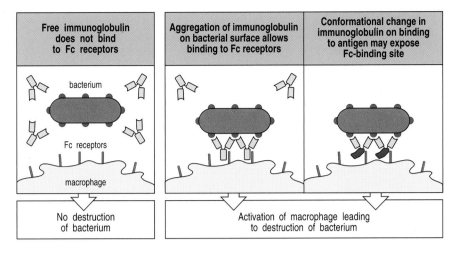

Free immunoglobulin does not bind to Fc receptors	Aggregation of immunoglobulin on bacterial surface allows binding to Fc receptors	Conformational change in immunoglobulin on binding to antigen may expose Fc-binding site
bacterium Fc receptors macrophage		
No destruction of bacterium	Activation of macrophage leading to destruction of bacterium	

Fig. 19.3 Bound antibody is distinguishable from free immunoglobulin by its state of aggregation and/or by conformational change. Free immunoglobulin molecules cannot bind Fc receptors. Antigen-bound immunoglobulin, however, can bind because several antibody molecules bound to the same surface bind to Fc receptors with high avidity. Some studies also suggest that antigen binding and aggregation induce a conformational change in the Fc portion of the immunoglobulin molecule, increasing its affinity for the Fc receptor. Both effects probably contribute to discrimination by Fc receptors between free and bound antibody.

2   *When an Rh-negative woman is compatible in the ABO system with her husband, as Mr and Mrs Waymarsh are, the risk of Rh alloimmunization is 16%. When they are ABO incompatible the risk falls to 7%. How do you explain this difference?*

If the mother and father are ABO identical, the fetus has a high likelihood of also being ABO compatible with its mother. If fetal blood enters the maternal circulation, it is likely to last much longer if the mother has no ABO alloantibodies to the fetal cells. For example, if the fetal red blood cells were type B, they would be quickly hemolyzed if the mother were red blood cell type A; she would have anti-B alloantibodies. Rapid destruction of the fetal red blood cells by hemolysis would impede alloimmunization of the mother.

3   *Why were Rh-negative red blood cells used for the intrauterine transfusion?*

Rh-negative cells were used because they cannot be hemolyzed by the IgG Rh antibodies that have crossed the placenta.

4   *Do you have concerns about administering Rhogam to women at 28 weeks of gestation?*

You shouldn't because the amount given (300 μg of IgG) is insufficient to cause the fetus harm. This amount of antibody raises the maternal titer in the indirect Coombs test to <1:4, a titer of Rh antibody that cannot cause significant hemolysis in the fetus. If the maternal titer is >1:4 she has been alloimmunized by her fetus.

5   *The serum of an Rh-negative woman who is pregnant gives a negative indirect Coombs test but her serum agglutinates Rh-positive cells suspended in saline. What is your interpretation of this phenomenon and what do you do about it?*

She has IgM anti-Rh antibodies; they agglutinate Rh-positive cells in saline, unlike IgG anti-Rh antibodies. As IgM antibodies do not cross the placenta you would have no immediate concern about her pregnancy. However, she should be tested repeatedly to be sure that she is not developing a positive Coombs test, signifying the presence of IgG antibodies.

Fig. 19.4 A rosette of Rh postive (Rh⁺) red cells. These are coated with anti-Rh antibodies, which adhere to the Fc receptors on a macrophage (central cell). The macrophage is pitting the red blood cells and destroying them. Photograph courtesy of J Jandl.

CASE 20

Autoimmune Hemolytic Anemia

Autoimmune disease triggered by infection.

Many clinical observations indicate that infection can trigger autoimmune disease, but in most instances the association between the infection and the autoimmunity is tenuous. There are various ways in which an infection could induce autoimmunity: disruption of a tissue barrier might expose a normally sequestered autoantigen; the infecting microorganism might act as an adjuvant; microbial antigens might bind to self proteins and act as haptens; and the microorganism might share cross-reactive antigens with the host (molecular mimicry) (see Fig. 17.6 and Question 3 in Case 17). The major histocompatibility antigens carried by an individual may also confer susceptibility to, or protection against, certain autoimmune diseases that can be incited by infection (Fig. 20.1).

Topics bearing on this case:
Coombs test
Humoral autoimmunity
Mechanisms for breaking tolerance

Fig. 20.1 Association of infection with autoimmune diseases. Several autoimmune diseases occur after specific infections and are presumably triggered by the infection. The case of post-streptococcal disease is best known. Most of these post-infection autoimmune diseases also show susceptibility linked to the MHC.

Associations of infection with immune-mediated tissue damage		
Infection	HLA association	Consequence
Group A *Streptococcus*	?	Rheumatic fever (carditis, polyarthritis)
Chlamydia trachomatis	HLA-B27	Reiter's syndrome (arthritis)
Shigella flexneri, Salmonella typhimurium, Salmonella enteritidis, Yersinia enterocolitica, Campylobacter jejuni	HLA-B27	Reactive arthritis
Borrelia burgdorferi	HLA-DR2, -DR4	Chronic arthritis in Lyme disease

There are only a few conditions in which the infection has been identified as the direct cause of the autoimmune disease. The bacterium *Streptococcus pyogenes*, for example, causes tonsillitis or pharyngitis (sore throat) or may cause a superficial skin infection called impetigo. Certain strains, however, lead to a type III autoimmune disease (caused by immune-complex formation; see also Case 23) of the renal glomeruli called post-streptococcal acute glomerulonephritis. These streptococcal strains are said to be nephritogenic (causing nephritis). About 3–4% of children infected with a nephritogenic strain of streptococcus will develop acute glomerulonephritis within a week or two of the onset of the streptococcal infection. What predisposes this subset of children to develop the complication is unknown.

A direct association of infection with autoimmune disease also occurs in patients with pneumonia caused by *Mycoplasma pneumoniae*, the case we shall discuss here. About 30% of patients with this infection develop a transient increase in serum antibody to a red blood cell antigen, and a small proportion of these patients develop hemolytic anemia, a type II autoimmune disease (see Fig. 17.1). In this case, the decrease in the number of red blood cells (anemia) results from their immunological destruction (hemolysis) as a result of the binding of an IgM autoantibody to a carbohydrate antigen on the red-cell surface. When the infection subsides, either spontaneously or after treatment, the autoimmune disorder disappears.

The IgM autoantibody agglutinates red blood cells at temperatures below 37°C. Thus the antibodies are called cold hemagglutinins or cold agglutinins, and the hemolytic disorder is known as cold agglutinin disease. Low titers of cold agglutinins (detectable at up to 1 in 30 dilution of serum) occur in healthy people. In about one-third of patients with mycoplasma infection there is a transient increase in the titer, but without symptoms; the rise usually goes unnoticed unless the patient happens to have a blood test, when clumping of the erythrocytes will be apparent.

The autoantibodies in the transient cold agglutinin syndrome resemble those in the chronic autoimmune disorder known as chronic cold agglutinin disease, which is of unknown etiology. This runs a protracted course in which the autoantibodies are characteristically oligoclonal, and lymphoproliferative disorders commonly develop. Thus chronic cold agglutinin disease behaves as a variant of an IgM gammopathy (an abnormal monoclonal or oligoclonal spike in the immunoglobulin electrophoretic pattern; see also Case 26) known as Waldenström's macroglobulinemia.

The case of Gwendolen Fairfax: the sudden onset of fever, cough, and anemia.

Gwendolen Fairfax was a healthy, unmarried 34-year-old who worked as a manager in a bank. She had never had any illness other than minor colds and the usual childhood infections. After developing a feverish cough her symptoms got progressively worse over the next few days and she decided to seek the advice of her physician, Dr Wilde.

34-year-old female with respiratory infection; anemia

Dr Wilde noted that Gwendolen was extremely pale; this was particularly noticeable in the palms of her hands, which were completely white. (Look at your palms; they have a healthy reddish pink color, unless your hemoglobin level is < 10 g dl^{-1}.) Her temperature was raised, at 38.5°C, and her respiratory rate was 30 per minute (normal 20). On listening to her chest, the physician heard scattered rhonchi (harsh breath sounds) at the bases of both lungs.

The physician ordered blood counts to be performed immediately. Gwendolen's hematocrit and hemoglobin level were low—hematocrit 26% (normal 38–46%) and hemoglobin 9.5 g dl^{-1} (normal 13–15 g dl^{-1}). Her white blood cell count was elevated at 11,300 μl^{-1}, with 70% neutrophils, 24% lymphocytes, 6% monocytes, and 1% eosinophils, revealing an increased absolute neutrophil count, and the platelet count was 180,000 μl^{-1}. The chest radiograph showed patchy infiltrates in both lower lungs (Fig. 20.2). He suspected a diagnosis of mycoplasma pneumonia (also known as atypical pneumonia because there was no complete opacity over a whole lobe of the lung such as typically occurs in pneumococal and other bacterial pneumonias). He advised Gwendolen to go to Reading Hospital.

Pneumonia with marked anemia; hemolytic anemia following mycoplasma infection.

In the hospital, a sputum sample proved negative for bacteria but contained abundant neutrophils. A blood sample was tested for the presence of antibodies against the red blood cells. In the presence of sodium citrate as an anticoagulant, the sample was chilled to 4°C; the red cells became clumped and the blood looked as though it had clotted. When warmed to 37°C, however, the red cells dispersed. Thus, the red cells were reversibly agglutinated in the cold. A blood sample was spun in a warmed centrifuge and the red cells and plasma separated. Gwendolen's red cells were tested using direct and indirect Coombs tests for the presence of anti-red cell IgG antibody (see Fig. 19.2); these tests proved negative. Her red cells, however, were strongly agglutinated by an antibody to the complement component C3.

When Gwendolen's plasma was incubated in the refrigerator with red cells from type O normal blood, the red cells were strongly agglutinated; when the same test was performed with red cells obtained from an umbilical cord in the delivery room of the hospital, agglutination was negligible.

Gwendolen was started on erythromycin by intravenous administration. In 3 days her symptoms had abated; she had no fever and her cough had greatly improved. Her hemoglobin had risen to 12 g dl^{-1}. She was discharged from the hospital, and advised to continue erythromycin orally and to avoid exposure to cold weather. When she returned to Dr Wilde's office two weeks later for a check-up, her fever had not returned, her cough had gone, and her hemoglobin was now a normal 14.5 g dl^{-1}.

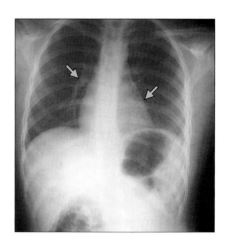

Fig. 20.2 Chest radiograph of a patient with mycoplasma pneumonia. Patchy opacities (arrowed) are evident in the lungs. Courtesy of T Griscom.

Fig. 20.3 Autoantibodies specific for the surface antigens of red blood cells can cause hemolytic anemia. Antibody-coated red blood cells (RBCs) can be destroyed in several ways. Cells coated with IgG autoantibodies are cleared predominantly by uptake by Fc receptor-bearing macrophages in the fixed mononuclear phagocytic system (left panel). Cells coated with IgM auto-antibodies and complement are cleared by macrophages bearing complement receptors CR1 and CR3 in the fixed mononuclear phagocytic system (not shown). The binding of certain rare autoantibodies that fix complement extremely efficiently causes the formation of the membrane-attack complement complex on the red cells, leading to intravascular hemolysis (right panel).

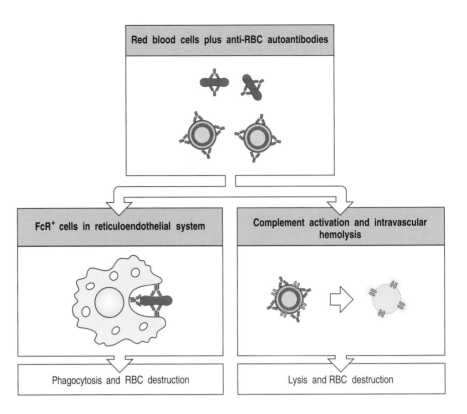

Autoimmune hemolytic anemia.

Gwendolen's autoimmune hemolytic anemia was transient, persisting only as long as the infection persisted. In contrast, autoantibodies to red blood cells in chronic autoimmune disorders of unknown origin tend to be persistent. Transient autoantibodies most frequently develop after non-bacterial infection. Autoimmune hemolytic anemia can also be induced by certain chemicals and drugs.

The autoantibodies in autoimmune hemolytic anemias are of the IgM or IgG class. They bring about hemolysis of the red blood cells either by complement fixation or adherence of the red cells to the Fcγ receptors on cells of the fixed mononuclear phagocytic system—principally in the spleen, but also in the liver and other organs (Fig. 20.3).

Anti-erythrocyte antibodies can differ in their physical properties. Some bind to the red blood cells at 37°C (warm autoantibodies); some at lower temperatures (cold hemagglutinins or hemolysins, like those in Gwendolen's case). In 1904, Donath and Landsteiner (who later got the Nobel Prize for the discovery of the ABO blood groups) reported antibodies in the blood of patients with syphilis that bound to human red blood cells at cold temperatures. These antibodies differ from cold agglutinins in that they cause the lysis of erythrocytes by complement when the blood is warmed to 37°C. These antibodies are still called Donath–Landsteiner or DL antibodies.

A transient increase in cold agglutinins is most frequently encountered during mycoplasma infection, infectious mononucleosis (the result of infection with the Epstein–Barr virus; see Case 12), and, rarely, during other viral infections. In mycoplasma pneumonia the cold autoantibodies react with a red-cell surface antigen called the I antigen. This is a branched carbohydrate structure that occurs on glycoproteins and glycolipids (Fig. 20.4). Cord blood cells express only small amounts of the I antigen and that is why Gwendolen's plasma did not agglutinate them. Cord erythrocytes express a related antigen, i (see Fig. 20.4); the I antigen is not fully expressed on red blood cells until 6 months of age.

It has not been possible to find with certainty a cross-reacting I antigen on *Mycoplasma pneumoniae*, which would provide a simple explanation for this autoimmune disease. Exactly how *M. pneumoniae* triggers the transient production of an autoantibody directed to red cells is not known, although there is a strong biochemical similarity between the attachment site (ligand) for the mycoplasma on the host cell surface and the red-cell antigen to which the autoantibody binds (Fig. 20.5).

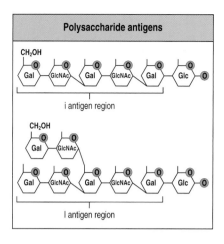

Fig. 20.4 Structure of the I and i antigen regions with which cold agglutinins react.

Discussion and questions.

1 The hemolytic anemia in Gwendolen Fairfax's case was not severe and it subsided rapidly. In similar cases the hemolysis can be extensive and the dramatic fall in hemoglobin can be life threatening. What emergency measure might be undertaken to stop severe hemolysis?

A plasma exchange (plasmapheresis). In this procedure the patient's blood is repeatedly removed 300 to 500 ml at a time and the blood is centrifuged (in this case at 37°C, at which temperature the IgM antibodies would be eluted from the red blood cells). The cells are re-suspended in normal plasma and infused back into the patient. This is a relatively efficient procedure for removing IgM antibodies because 70% of IgM is in the plasma compartment and only 30% of IgM is in the extravascular compartment. In contrast (as we learned in Case 2), only 50% of IgG is in the vascular compartment and 50% is in the extravascular space.

2 Autoimmune hemolytic anemia is caused, in general, by autoantibodies of the IgG and IgM classes. Could antibodies of other classes such as IgA, IgD, and IgE cause autoimmune hemolytic anemia?

They usually do not cause anemia because they do not fix complement (C1q) and there are no Fc receptors for these immunoglobulin classes on cells of the macrophage lineage, which bear only Fcγ receptors. Fcε receptors are expressed on mast cells and B cells, and their engagement would not result in the destruction of red blood cells.

3 In the direct and indirect Coombs test (see Case 19), agglutination of the antibody-coated red blood cells is brought about by an antibody to human IgG that crosslinks the IgG on the red-cell surface and causes the cells to agglutinate. In the so-called 'non-gamma' Coombs test an antibody to C3 or C4 is used to agglutinate red blood cells coated with an IgM antibody that has fixed complement. Why is it important to have an antibody to the C3d or C4d portions of C3 or C4 to carry out the test?

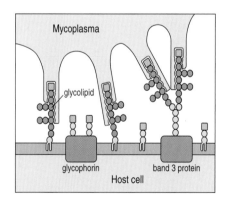

Fig. 20.5 *Mycoplasma pneumoniae* interacts with long carbohydrate chains of I antigen type on the human red cell membrane. These chains include the carbohydrates of band 3 protein and certain glycolipids, but not glycolipids and glyco-proteins with short carbohydrate chains. Sequences of sugars that act as I antigen are shown as orange circles; the orange squares depict sialic acid.

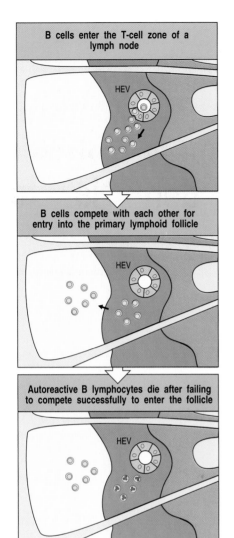

Fig. 20.6 Autoreactive B cells do not compete effectively to enter primary lymphoid follicles in peripheral lymphoid tissue. In the top panel B cells are seen entering the T-cell zone of a lymph node through high endothelial venules (HEVs). Those with reactivity to foreign antigens are shown in yellow and autoreactive cells are gray. The autoreactive cells fail to compete with B cells specific for foreign antigens for exit from the T-cell zone and entry into primary follicles (center panel). This is because B cells reactive to foreign antigens receive signals from antigen-specific T cells which promote their activation and survival. In contrast, the autoreactive B cells fail to receive survival signals and undergo apoptosis in the T-cell zone (bottom panel).

The internal thiol ester that binds covalently to the IgM autoantibody is situated in the C3d and C4d fragments. The complement components C3b and C4b that are bound to the autoantibody may be digested by Factor I to C3c + C3d and C4c + C4d. C3c and C4c would be released from the antigen:antibody complex and thus antibodies against them would not agglutinate the red blood cells.

4   *Most normal individuals have a low titer (up to 1:30) of IgM anti-I cold agglutinins in their blood. As mentioned above, the I antigen is little expressed until 6 months of age. We must develop peripheral B-cell tolerance to this antigen in infancy. How does this happen? How might the mycoplasma infection break this tolerance?*

There is almost certainly no T-cell antigen receptor for this carbohydrate antigen, so that help to move the B cells into follicles would not occur and specific B cells would undergo apoptosis in the T-cell zone (Fig. 20.6). Alternatively, the B cells might be rendered anergic by soluble antigen or undergo programmed cell death through interaction of Fas and Fas ligand.

Although the precise mechanism of autoimmunization awaits elucidation, the molecular characterization of the autoantigen I and of the host cell adherence receptors for *Mycoplasma pneumoniae* have indicated collectively that the disorder has its origin in the interaction of the infective agent with carbohydrate attachment sites on host cells. The mycoplasma adheres to the ciliated bronchial epithelium, and also to red cells and a variety of other cells via ligands that consist of long carbohydrate chains of I antigen type that are capped with sialic acid (see Fig. 20.5). When bound to the carbohydrate chain that contains the I-antigen sequence, the mycoplasma may act as a carrier and the I antigen may act as a hapten.

CASE 21 | **A Kidney Graft for Complications of Autoimmune Insulin-Dependent Diabetes Mellitus**

Mechanisms of graft rejection and HLA associations in autoimmunity.

The transplantation of solid organs is becoming increasingly common as surgeons have become more skilled in transplantation techniques and immunosuppressive drugs have become more effective. Only the supply of donor organs seems to limit this growing surgical practice. The first successful solid organ transplant was a kidney graft from one identical twin brother to the other at the Peter Bent Brigham Hospital in Boston in the mid-1950s. Since then, solid organ transplantation has expanded to successful transplants of liver, heart, lungs, and pancreas.

Ideally, solid organ transplantation should be performed between histoidentical donors and recipients but this, for all practical purposes, is not possible. Perhaps only 2 in 10,000 unrelated individuals are histocompatible at the major histocompatibility locus (MHC). Among sibling donor recipient pairs, 1 in 4 should be histoidentical at the MHC but there are further problems with compatibility of ABO blood groups and minor histocompatibility loci. It is now clear that, in solid organ transplantation, absolute matching at the MHC loci offers only a small incremental advantage for survival of the transplanted organ. This realization has led to the widespread use of unrelated cadaveric organ donors.

The first risk to be considered is the problem of hyperacute rejection. It is important that the recipient matches the donor at the ABO blood group locus lest antibodies in the recipient react to A or B blood group substances in the endothelium of the graft. Such antigen–antibody interactions in the small blood vessels of the graft induce clotting in these small blood vessels; the graft becomes hemorrhagic and quickly dies. It is also very important to ascertain that the recipient has no antibodies to the cell-surface antigens of donor white cells. Such antibodies might be present, for example, if the recipient has had a blood transfusion on an earlier occasion: leukocytes are present in most transfused blood, and prior sensitization to leukocyte antigens of the donor can cause hyperacute rejection of the graft within a few hours. Hyperacute rejection is impossible to treat and when it occurs the graft cannot be saved.

Topics bearing on this case:

Tissue matched for organ transplants

Transplant rejection and immunosuppressive drugs

HLA-disease association

T-cell mediated autoimmune disease

Mechanisms of action of immunosuppressive drugs

Between 7 and 15 days after solid organ transplantation the host T cells will mount an immune response to the histocompatibility antigens of the donor. There is usually a general inflammatory reaction, with massive lymphocyte infiltration culminating in infiltration by CD8 cytotoxic T cells and first-set graft rejection. Unlike hyperacute rejection, first-set graft rejection can be treated satisfactorily, and immunosuppressive drugs such as azathioprine, prednisone, cyclosporin A and FK506 allow long-term survival of grafts.

The case of Christopher Goodwood: a saga of organ rejection and acceptance.

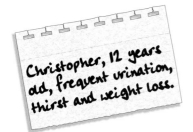

Christopher, 12 years old, frequent urination, thirst and weight loss.

Hypertension, proteinuria and elevated serum creatinine.

Christopher Goodwood was the only child born to his parents. His growth and development were normal and he excelled at school. When he was 12 years old he began to lose weight and developed such an enormous thirst that he was consuming over 3 liters of liquid a day. The thirst and excess fluid consumption was accompanied by frequent urination. He was taken to the family doctor who diagnosed diabetes mellitus. It turned out that several relatives of Christopher's father had also developed juvenile onset diabetes.

Christopher was started on daily insulin injections but his blood sugar was always difficult to control. Despite this handicap, Christopher continued to do well in school and he eventually entered college as an engineering student. Shortly after he graduated he found a job with a civil engineering firm and was married.

When he was 35 years old, during a routine annual physical examination, his physician found that his blood pressure was elevated and that his urine tested positive for the presence of protein. A serum creatinine test was obtained and found to be 7.5 mg dl^{-1} (normal <1.0 mg dl^{-1}). On the basis of these results, Christopher's physician suspected that he had developed renal complications of diabetes.

The physician prescribed drugs to control the high blood pressure and recommended that a kidney biopsy be performed. The biopsy, which entails the removal of a small sample of tissue, revealed extensive diabetic glomerulosclerosis (scarring of the glomeruli by fibrous tissue) (Fig. 21.1), showing that Christopher was in imminent danger of kidney failure. He was therefore scheduled for hemodialysis (passing blood over a semipermeable membrane to remove waste products such as urea) twice a week. He was also put on a waiting list for a cadaveric kidney transplant.

Christopher was blood group B, Rh-positive. His HLA type was A2,24; B50,51; DR3,4. After a waiting period of 6 months during which he had to continue on hemodialysis, a cadaveric kidney was found from a 20-year-old man who was fatally injured in a motorcycle accident. The donor was blood type B, Rh-positive. Only HLA-A and -B typing were available from the donor. He was HLA-A 2,11; -B7,35. Christopher's serum was tested for antibodies against the white blood cells of the potential donor and was found to be non-reactive.

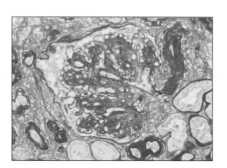

Fig. 21.1 Glomerulus from a diabetic with glomerulsclerosis. Photograph courtesy of R Colvin and A Perez-Atayde.

Christopher was admitted to the hospital and, on the following morning, his own kidneys were removed by a surgeon and replaced with the donor kidney. During the procedure Christopher was given 100 mg methylprednisolone and 40 mg of anti-ICAM monoclonal antibody. During the first day after his operation the transplanted kidney put out 200 ml of urine. This is only about one-fifth the normal output for a healthy person but usual for a newly transplanted kidney and showed that the organ was functioning.

Christopher was given 450 mg cyclosporin, 100 mg azathioprine, 160 mg methylprednisolone and a further 40 mg of anti-ICAM antibodies. The serum creatinine returned to normal during the next 5 days. After the fifth day anti-ICAM antibodies were stopped and Christopher was continued on 100 mg azathioprine, 400 mg cyclosporin and 20 mg prednisone. Christopher was discharged from the hospital 10 days after the transplant was performed. His blood pressure and serum creatinine were normal. The transplant appeared normal by ultrasonography.

One week after discharge from the hospital Christopher noticed that his urine output had fallen to about half normal. When he went back to the hospital, his creatinine proved to have risen to 2.5 mg dl^{-1} and the graft was found to be enlarged and tender. A renal (kidney) biopsy revealed the presence of many lymphocytes in the transplant (Fig. 21.2) compared with a normal kidney, indicating that the kidney was undergoing rejection.

His immunosuppressive therapy was increased to 500 mg methylprednisolone intravenously, the cyclosporin was continued, and intravenous injections of 5 mg of a monoclonal antibody to CD3 were given for 10 days. His peripheral blood after this regimen contained 15% CD3^{+} cells. This immunosuppressive regimen successfully suppressed the symptoms and Christopher continued to do well for another 6 weeks.

Six weeks later, however, while he was recuperating at his parents' home, Christopher developed fever and a severe cough that produced yellow, thick sputum. His physician was called and, on examining Christopher, heard crackles and wheezes over both lungs. A sample of Christopher's sputum was tested for *Mycobacterium tuberculosis*, which could not be detected, and for Gram-negative and Gram-positive bacteria, which were present but sparse; this is normal. A silver stain for *Pneumocystis carinii* was also negative. The fungus *Aspergillus fumigatus* was cultured from all his sputum specimens. He was treated with the fungicide amphotericin B, and recovered.

Within a year of the transplant Christopher returned to his civil engineering job. His blood pressure remains normal, his serum creatinine is <1.0 mg dl^{-1} and he continues to take low doses of azathioprine, prednisone and cyclosporin.

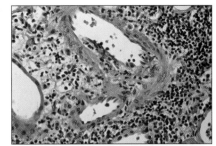
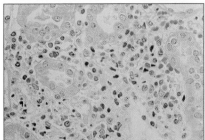
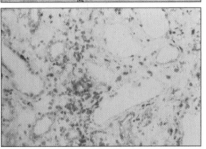

Fig. 21.2 First-set rejection in a kidney graft. Top panel shows lymphocytes around an arteriole in the kidney. The middle panel shows lymphocytes surrounding the renal tubules of a kidney undergoing rejection and the bottom panel shows the staining of lymphocytes with anti-CD3 in the same section.

Insulin-dependent diabetes mellitus and kidney grafting.

Insulin-dependent diabetes mellitus (IDDM) is a T-cell mediated autoimmune disease (Fig. 21.3). Like many other autoimmune diseases, it occurs with a markedly higher frequency in individuals with certain specific MHC class II alleles (Fig. 21.4). The reasons for this association are obscure. Although the MHC alleles seem to predispose to the development of IDDM, it is clear that other factors also have a role. For example, if IDDM develops in one of two identical twins, the chances are only one in three that the other will develop it as well. This suggests that some environmental factor determines whether the disease is actually triggered.

Fig. 21.3 Common T-cell mediated autoimmune diseases.

T-cell mediated autoimmune disease		
Syndrome	Autoantigen	Consequences
Insulin-dependent diabetes mellitus	Pancreatic β-cell antigen	β-cell destruction
Rheumatoid arthritis	Unknown synovial joint antigen	Joint inflammation and destruction
Experimental autoimmune encephalomyelitis (EAE), multiple sclerosis	Myelin basic protein, proteolipid protein	Brain invasion by CD4 T cells, paralysis

IDDM is also known as juvenile onset diabetes because it usually appears in childhood or early adolescence. At the beginning of the disease, the islets of Langerhans in the pancreas become infiltrated with lymphocytes, which selectively destroy the beta cells of the islets (Fig. 21.5). The antigen that provokes the autoimmune attack on the beta cells is not known and has been difficult to identify. An excellent animal model of IDDM is provided by NOD (non-obese diabetic) mice, in which, as in humans, a specific MHC class II allele is associated with the development of diabetes. A very high proportion of these mice develop diabetes but not all, which, as in humans, indicates an environmental factor.

It has been possible to retard the progression of diabetes in NOD mice by administering immunosuppressive agents or by depleting these mice of T cells. It has also been possible to transfer the disease with T cells. These observations have led to various immunosuppressive therapeutic trials in humans with IDDM but success has not yet been achieved convincingly.

The case of Christopher Goodwood illustrates one of the long-term complications of poorly controlled IDDM (that is, IDDM in which blood sugar is poorly regulated): namely diabetic glomerulosclerosis. The kidneys are damaged irreversibly by this complication of IDDM, renal failure ensues and the only available therapy, other than hemodialysis, is renal transplantation. At best, the kidney recipient may expect perhaps 20 years of useful life from the graft before it too develops glomerulosclerosis. In practice, ill-matched grafts usually do not survive that long, and may last no longer than 5 years.

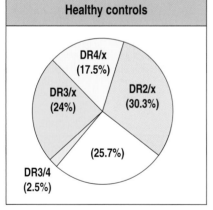

Healthy controls

DR4/x (17.5%)
DR3/x (24%)
DR2/x (30.3%)
(25.7%)
DR3/4 (2.5%)

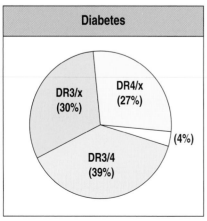

Diabetes

DR3/x (30%)
DR4/x (27%)
(4%)
DR3/4 (39%)

Fig. 21.4 Population studies show association of susceptibility to insulin-dependent diabetes mellitus (IDDM) with HLA genotype. The HLA genotypes (determined by serotyping) of diabetic patients (bottom panel) are not representative of those found in the population (top panel). Almost all diabetic patients express HLA-DR3 and/or HLA-DR4, and HLA-DR3/DR4 heterozygosity is greatly over-represented among diabetics compared with controls. These alleles are linked tightly to HLA-DQ alleles that confer susceptibility to IDDM. By contrast, HLA-DR2 protects against the development of IDDM and is only found extremely rarely in diabetic patients. The small letter x represents any allele other than DR2, DR3, or DR4.

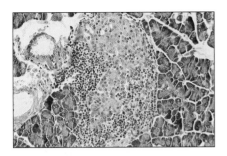 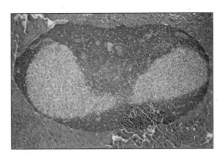

Fig. 21.5 Pancreatic beta cells with lymphocytic infiltration. The left panel is an islet from a 6-week-old non-obese diabetic (NOD) mouse showing the invasions of the islet with lymphocytes. The right panel is the islet of an NOD mouse treated with complete Freund's adjuvant. This causes massive and dramatic lymphocyte invasion of the islet. Photographs courtesy of J Katz (left) and O Kanagawa (right).

Discussion and questions.

1 Why did Christopher develop an opportunistic infection with Aspergillus fumigatus?

These infections are normally controlled by macrophages activated by T cells. Christopher's cell-mediated immunity was suppressed by azathioprine, prednisolone, prednisone, and cyclosporin A.

2 How does cyclosporin A work?

It prevents the synthesis of interleukin-2 and other cytokines that are essential to the clonal expansion of T cells and also the activation of their effector functions. Cyclosporin A acts by inhibiting the activation of the cytokine genes by the transcriptional activator NF-AT. NF-AT is a cytoplasmic phosphoprotein that is activated when it is dephosphorylated by calcineurin. Dephosphorylation enables NF-AT to translocate to the nucleus, where it activates the synthesis of cytokines. Cyclosporin A inhibits the dephosphorylation of NF-AT by calcineurin.

3 When the graft was undergoing first-set rejection, it became swollen and tender. How do you explain this? What would second-set rejection be?

The graft became infiltrated with lymphocytes that caused a general inflammatory response, as well as a specific attack on the cells in the graft by cytotoxic CD8 T cells. The graft became swollen with edema fluid and inflammatory cells. The renal capsule was stretched by the edema fluid and cellular infiltrate, and this induced pain.

The donor was HLA-A2,11 and HLA-B7,35. If Christopher received another graft with the markers HLA-A11, or HLA-B7 or 35 he would reject such a graft more rapidly because he would have been sensitized to these HLA markers. This is second-set rejection.

4 Why was it considered more important to match the potential donor's blood group with Christopher's blood group and to ascertain that Christopher had no cytotoxic antibodies to the donor than to worry about complete HLA matching?

To avoid hyperacute rejection, for which there is no therapy. If these events had occurred, a valuable kidney graft would have been wasted. On the other hand, HLA mismatching would lead to problems with first-set rejection, which could be managed with immunosuppressive therapy.

[5] Why did the surgeon choose to give Christopher anti-ICAM monoclonal antibodies and subsequently to treat impending rejection with anti-CD3?

Anti-ICAM interferes with the interaction of LFA-1 on lymphocytes and other leukocytes with ICAM-1 expressed on endothelial cells. Thus it would impede the emigration of white cells from the circulation into the graft (see Case 13). When Christopher's kidney graft was threatened with rejection, it became important to get rid of the T cells mustering to attack it. Antibody to CD3 would remove T cells from the circulation and lower the total body T cells as it apparently did in this case (only 15% of Christopher's peripheral blood lymphocytes (PBL) reacted with anti-CD3, compared with a normal value of 75–80%).

[6] Several members of Christopher's father's family had also developed juvenile onset diabetes. Can you suggest why?

Juvenile onset diabetes is strongly associated with certain MHC class II alleles (see Fig. 21.3). It is likely that Christopher and his father's diabetic relatives share one or more of these alleles.

[7] Christopher is HLA-DR3. This MHC type makes him four times more susceptible to IDDM (as well as to other autoimmune diseases) as the average person. But he is also HLA-DR4, and the combination of HLA-DR3 and HLA-DR4 makes him 50 times more susceptible. What HLA-DR type in place of HLA-DR4 would have lowered his risk of acquiring IDDM?

HLA-DR2 is least frequently encountered in people with IDDM and is therefore associated with a lower risk.

[8] Would you expect to find the mechanism of Christopher's disease susceptibility by looking at the properties of molecules belonging to the HLA-DR3 and HLA-DR4 types?

Not necessarily. The fact that HLA-DR3 and HLA-DR4 are associated with IDDM does not necessarily mean that they are involved directly in the mechanism of susceptibility. The HLA-DR3 and HLA-DR4 genes could simply be very close to some other gene that is therefore usually inherited with them and is the real reason for the IDDM susceptibility. This problem, which is known as linkage disequilibrium, applies to any genetically defined disease association. In fact, the inheritance of susceptibility to IDDM is now known to be more closely associated with the inheritance of specific alleles of HLA-DQ genes, which are adjacent to the HLA-DR genes in the MHC (see Fig. 4.3). Molecules belonging to the HLA-DQ type are difficult to type by using antibodies, and this association has only become clear since it became possible to look directly at the DNA encoding them. Most of the information we have on susceptibility to IDDM therefore comes from patients who were typed only for HLA-DR alleles.

9   *How might specific HLA-D alleles confer susceptibility to IDDM, and conversely how might other alleles lower this risk?*

HLA-DR and HLA-DQ are class II MHC molecules that present peptide antigens to CD4 T cells. You might therefore expect that specific alleles of these molecules are particularly efficient at presenting peptides derived from pancreatic beta cells and this could explain why T cells become activated against these cells in IDDM. But it is, in fact, not at all obvious how this would work, and the mechanism of disease susceptibility in IDDM is an outstanding issue in research immunology. The problem is that pancreatic beta cells themselves cannot present their peptides to CD4 T cells because they do not express MHC class II molecules (and even if they did, MHC class II molecules do not present antigens generated endogenously). The direct target of the autoimmune attack by T cells on pancreatic beta cells is therefore not known, and neither is the connection between that target and the MHC class II molecule or molecules associated with IDDM.

The mechanism of protection by HLA alleles associated with low risk is also unclear. One obvious possibility is that these HLA molecules cannot bind antigenic pancreatic beta-cell peptides. However, this cannot be the explanation, because the protective alleles actually prevent the development of IDDM in individuals with strong susceptibility alleles. This has been demonstrated conclusively in NOD mice, into which genes for protective MHC molecules can be introduced, and the resulting transgenic animals can be shown to be protected from developing IDDM. The explanation for these dominant protective effects is again unknown.

10   *How might environmental factors contribute to the development of IDDM?*

Perhaps not surprisingly, the proportion of NOD mice developing diabetes has been found to vary depending on how carefully the animal houses are kept free from infection. Perhaps more surprisingly, high rates of infection are associated with lower rates of IDDM. This is another unexplained but important phenomenon: there are other autoimmune diseases in which decreased susceptibility is associated with infectious disease. For example, systemic lupus erythematosus, which we discuss in Case 23, is never seen in black Africans in Africa where parasitic disease is endemic. When African blacks migrate to temperate countries where there is no parasitic disease, however, the incidence of lupus quickly approaches that seen in the native Caucasian population. In other autoimmune diseases, by contrast, infection clearly increases susceptibility (see Case 17).

CASE 22 | Pemphigus Vulgaris

Autoimmune attack on the integrity of the skin.

Autoimmune disease results when the adaptive immune response is directed against self antigens. The particular pathology of the disease depends on the nature of the self antigen and the type of immune response that is mounted against it. In the type II autoimmune diseases (see Fig. 17.1) the pathology is caused by autoantibodies that interact with self antigens in the extracellular matrix or on cell surfaces. In myasthenia gravis, for example, autoantibodies against the acetylcholine receptor on skeletal muscle block its function at the neuromuscular junction and cause paralysis (see Case 17), whereas in Graves' disease an autoantibody to the receptor for thyroid-stimulating hormone acts as an agonist, stimulating receptor activity and causing hyperthyroidism. In the autoimmune disease pemphigus vulgaris, the self-reactive agent is an autoantibody against a structural protein of the epidermal cells of the skin. Its actions result in the skin cells coming apart from each other; the affected skin blisters and is destroyed.

Pemphigus is derived from the Latin word for blister and vulgaris means common or ordinary. Patients with this disease have an autoantibody to desmoglein-3, which is a protein component of the desmosome—one of the intercellular junctions that link skin cells and other epithelial cells tightly to each other. Desmoglein-3 is a member of the cadherin family of cell adhesion molecules, proteins that effect intercellular adhesion in a calcium-dependent manner. Disruption of desmoglein-3 causes blisters to form in the skin and on the mucous membranes; extensive sloughing of the skin may ensue (Fig. 22.1).

Topics bearing on this case:
Humoral autoimmunity
HLA associations with disease
Peptide binding by MHC class II antigens

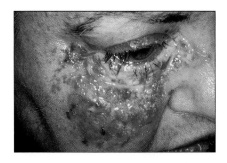

Fig. 22.1 Lesions of pemphigus vulgaris that have coalesced to produce a large plaque. The plaque has a moist surface that is oozing and has crusts of dried serum from blister fluid. Photograph courtesy of Razzaque Ahmed.

Persistent skin sore and ulceration of mouth; resistant to usual therapies; do biopsy

Diagnosis of pemphigus vulgaris; start immuno-suppressive therapy

Condition not responding to steroids; try cyclo-phosphamide

Evidence that pemphigus vulgaris is caused by an autoantibody comes from the observation that infants born to mothers with the disease have a transient period of skin blistering during the neonatal period. There is also a good animal model in which human IgG from patients with pemphigus injected into neonatal mice reproduces pemphigus.

The case of Arthur Sammler, Esq.: a hoarse lawyer with a sore mouth and blistering skin.

Arthur Sammler is a lawyer of Ashkenazi Jewish descent. He was 55 years of age when an irritation in his throat became persistent enough for him to consult a physician. The physician suggested that he gargle with warm saline, which seemed to help. A month later, however, after a long day in court he became very hoarse and the hoarseness persisted for a week. A week later he noticed a 'sore' on his right cheek. He told the physician he was under considerable stress over a custody case, and the physician diagnosed a herpes simplex infection and advised him to take oral Acyclovir. The lesion did not improve.

At a more thorough examination two weeks later, numerous erosions and ulcers in the mucosa of his mouth and gums were also revealed. These were thought to result from a yeast infection with *Candida albicans* and he was treated with a Mycostatin (nystatin) mouth wash.

However, new lesions soon appeared on the palate, uvula, and tongue, and Mr Sammler returned to the physician. This time, a biopsy of the lesions was taken by an oral surgeon. While awaiting the results of the biopsy Mr Sammler developed erosions of the skin on his scalp and neck.

Microscopic examination of the biopsy revealed disruption of the epidermal layer of the skin characteristic of pemphigus vulgaris (Fig. 22.2). Mr Sammler was sent to a dermatologist, who took another biopsy for immunofluorescence studies; blood samples were taken for analysis. The immunofluorescence studies revealed deposits of IgG in an intercellular pattern within the epidermis (Fig. 22.3). His serum contained an antibody to desmoglein-3 (known commonly as 'pemphigus antibody') in a titer of 1:640.

Mr Sammler was treated with large doses of prednisone, 120 mg per day. After three weeks of this therapy, during which time no improvement occurred, new lesions appeared on his back and scalp. The dose of prednisone was increased to 180 mg per day. Two weeks later he developed a persistent cough, fever and chest pain. A chest radiograph during an emergency room visit suggested that he had developed interstitial pneumonia due to *Pneumocystis carinii*. He was admitted to hospital and treated for this infection.

It was found that his pemphigus antibody titer had risen to 1:1280. He was started on cyclophosphamide at 150 mg per day, and was told to decrease his prednisone dose by 10 mg every 10 days and to rinse his mouth daily with hydrogen peroxide and elixir of Decadron (a corticosteroid). New lesions kept appearing so the cyclophosphamide was increased to 200 mg per day.

Over the next 8 weeks the pemphigus antibody titer decreased to 1:320; no new lesions appeared, and the older ones healed. The cyclophosphamide dose was decreased over an 8-month period and he remained symptom-free after cyclophosphamide was stopped. In the year since then, he has noticed the occasional new lesion after damage to his skin and these new lesions are promptly injected with a corticosteroid. His pemphigus antibody titer has remained at 1:80 during this asymptomatic period.

Responding to cyclophosphamide; decrease dose gradually

Pemphigus vulgaris.

The case of Arthur Sammler illustrates how an autoimmune disease that was frequently fatal in the past can now be controlled by immunosuppressive drugs. The epidermis, with its cells held tightly together, forms an important barrier to the entry of pathogens into the body. Extensive blistering and sloughing of skin destroy this barrier and, in the past, often resulted in bloodstream infection (septicemia) with *Staphylococcus aureus*, a common contaminant of human skin.

The mechanism of autoimmune tissue destruction seems to be somewhat unusual in pemphigus vulgaris. In some autoimmune diseases, the autoantibodies cause tissue destruction by stimulating the complement system, but the autoantibody in pemphigus vulgaris is of the IgG4 subclass and does not fix complement. Some other mechanism must therefore be invoked to explain the disruption of desmoglein-3 adhesion. It is thought that the binding of the autoantibody to its antigen in some way causes upregulation of serine proteinase activity on the surface of the epidermal cells and that this results in the proteolytic digestion of desmoglein-3. The particular enzyme that digests the desmoglein-3 is as yet unknown.

Pemphigus vulgaris is encountered most frequently in Ashkenazi Jews and has a strong association with an HLA haplotype found mainly in that ethnic group. This HLA haplotype includes HLA-DR4 and HLA-DQ3, which are present in virtually all Ashkenazi Jews with pemphigus vulgaris. Patients are usually heterozygous for this haplotype and only in rare cases homozygous. Unaffected relatives of pemphigus patients who bear the same HLA haplotype frequently also have antibodies to desmoglein-3. However, these antibodies are of the IgG1 subclass and do not cause disease. In other ethnic groups, pemphigus vulgaris is usually associated with HLA-DR14 and HLA-DQ5.

A molecular explanation has been discovered for the association of pemphigus vulgaris with HLA-DR4. Each MHC class II variant can bind a subset of all possible peptides; the actual peptides that can be bound by any given MHC molecule are determined by the particular pattern of amino-acid binding sites in the peptide-binding groove of the MHC molecule (Fig. 22.4). The β chain of the MHC class II molecule DR is designated DRB1; 22 known subtypes of this chain have been identified by DNA typing among those bearing DR4 (DRB1*04). Only one of these variants—DRB1*0402—is associated with pemphigus vulgaris. It differs from the other 21 variants in that the amino acid residue at position 71, which lies in one of the binding pockets in the peptide-binding groove of the MHC molecule, is negatively charged (glutamic acid) whereas all the other variants have positively charged amino acid residues (lysine or arginine) at this position. Only DRB1*0402 will bind the antigenic peptide derived from desmoglein-3 and present it to T lymphocytes.

Fig. 22.2 Histopathology of an early skin lesin of pemphigus vulgaris. Stained with hematoxylin and eosin. An epidermal, suprabasal vesicle (blister) is clearly seen. Photograph courtesy of Razzaque Ahmed.

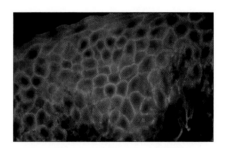

Fig. 22.3 Direct immunofluorescence study of tissue around a pemphigus vulgaris lesion. The normal-appearing tissue around the lesion contains deposits of IgG, other immunoglobulins, and complement components in the intercellular spaces of the entire stratum malpighii of the epidermis. Photograph courtesy of Razzaque Ahmed.

Fig. 22.4 Peptides bind to MHC class II molecules by interactions along the length of the binding groove. A peptide (yellow; shown as the peptide backbone only, with the amino terminus to the left and the carboxy terminus to the right), is bound by an MHC class II molecule through a series of hydrogen bonds (dotted blue lines) that are distributed along the length of the peptide. The hydrogen bonds towards the amino terminus of the peptide are made with the backbone of the class II polypeptide chain, whereas throughout the peptide's length bonds are made with residues that are highly conserved in MHC class II molecules. The side chains of these residues are shown in gray on the ribbon diagram of the MHC class II binding groove. Most DR4 alleles have a positively charged arginine (shown here in purple) or lysine at position 71 of the β chain. The DRB1*0402 allele that is linked to susceptibility to pemphigus has a negatively charged glutamic acid residue instead.

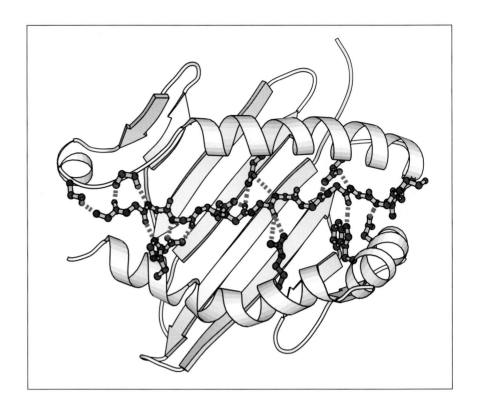

Discussion and questions.

1   The intravenous administration of gamma globulin frequently improves autoimmune disease. It has been observed that the administration of gamma globulin to patients with pemphigus vulgaris significantly lowers the titer of IgG4 anti-desmoglein-3 and improves the clinical course of the disease. How do you explain this effect?

The monomeric IgG in the gamma globulin administered intravenously binds to high-affinity Fc receptors on macrophages and causes the release of immunosuppressive cytokines such as transforming growth factor-β (TGF-β), interleukin (IL)-10 and the IL-1 receptor antagonist. In patients with pemphigus vulgaris, clinical improvement after intravenously administered gamma globulin is noted in about 6 weeks and decreased antibody titers are found after 6 months.

*2   Relatives of patients with pemphigus vulgaris who share the DRB1*0402 MHC class II type have IgG1 antibodies to desmoglein-3. These antibodies fix complement but do not cause disease. Can you speculate why this antibody does not cause disease whereas the IgG4 antibody causes disease?*

The reasons for this difference are not understood. We know that the IgG1 antibody reacts with a different epitope of desmoglein-3 than does the IgG4 antibody, but this in itself does not provide an explanation. IgG1 fixes complement, and thus we must infer that the proteinases generated by complement activation do not digest desmoglein-3.

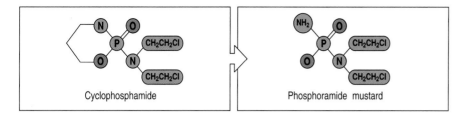

Fig. 22.5 The structure and metabolism of the cytotoxic drug cyclophosphamide. Cyclophosphamide is administered as a stable 'pro-drug' which is transformed enzymatically in the body to phosphoramidate mustard, a powerful and unstable DNA alkylating agent.

3 *What might explain the difference between unaffected people who make only the IgG1 antibody to desmoglein-3 and the patients who make IgG4 antibody? Can you design an experiment to test your hypothesis?*

Isotype switching is required to make the IgG4 antibodies in addition to or in lieu of the IgG1 antibodies. Switching to IgG4 is stimulated by the cytokine IL-4 released by activated T_H2 cells. If we take T cells from asymptomatic individuals who make IgG1 antibodies and T cells from patients who make IgG4 antibodies and stimulate them with desmoglein-3 we might find that the patients' T cells make more IL-4 (and thus undergo more isotype switching) than those of the asymptomatic individuals. Higher levels of IL-4 production in patients' T cells are indeed found in such an experiment.

4 *We have seen cyclophosphamide used to eradicate bone marrow cells in Cases 13 and 18. Why is it useful in the treatment of pemphigus vulgaris?*

Cyclophosphamide (Fig. 22.5) is an alkylating agent that interferes with DNA synthesis and therefore stops cell division. Although it has many bad side effects such as anemia, thrombocytopenia and hair loss, it is effective in halting lymphocyte cell division, and thus suppresses immune reactions.

5 *Why did Mr Sammler develop pneumonia caused by Pneumocystis carinii during his initial treatment for pemphigus vulgaris?*

Pneumocystis carinii is an opportunistic pathogen that is a frequent cause of pneumonia in immunosuppressed patients. In this case the high-dose corticosteroid treatment suppressed T-cell functions and trafficking, causing increased susceptibility to this potentially serious infectious agent.

CASE 23 | Systemic Lupus Erythematosus

A disease caused by immune complexes.

Immune complexes are produced whenever there is an antibody response to a soluble antigen. As the immune response progresses, larger immune complexes are formed that trigger the activation of complement, activated components of which bind to the complexes. These are then efficiently cleared by binding to complement receptor 1 (CR1) on erythrocytes, which convey the immune complexes to the liver and spleen. There the complexes are removed from the red cell surface by Kupffer cells and other phagocytes (Fig. 23.1) and ingested via a variety of complement and Fc receptors on the cells lining the sinusoids of the hepatic and splenic circulation. When antigen is released repeatedly so as to sustain the formation of small immune complexes, these complexes tend to be trapped in the small blood vessels of the renal glomerulus and synovial tissue of the joints.

Topics bearing on this case:

Clearance of immune complexes by complement

Immune-complex disease

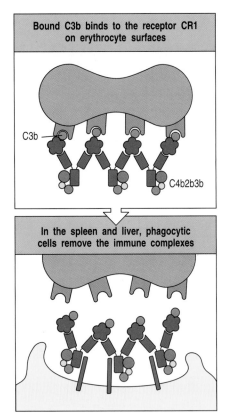

Bound C3b binds to the receptor CR1 on erythrocyte surfaces

C3b

C4b2b3b

In the spleen and liver, phagocytic cells remove the immune complexes

Fig. 23.1 Immune complexes are cleared from the circulation by binding to complement and Fc receptors. Immune complexes activate C3 in the serum, and bind activated complement components C3b, C4b, and C2b. C3b binds to complement receptors on erythrocytes, which transport the immune complexes to the spleen and liver, where complement receptors and Fc receptors on phagocytic cells bind to complement components and to the Fc portion of antibodies and are thereby stimulated to engulf the complexes and degrade them.

The most common immune-complex diseases are listed in Fig. 23.2. In subacute bacterial endocarditis, bacteria reside for a protracted period on the valves of the heart. The antibody response to the prolonged presence of the bacteria becomes intense and immune complexes are formed with the bacterial antigens. The immunoglobulin in the immune complexes provokes the formation of anti-IgG antibodies or rheumatoid factor. These complexes are trapped in the renal glomeruli and cause glomerulonephritis. In a similar fashion, hepatitis virus in the liver can become a chronic infection that provokes a marked antibody response and the formation of rheumatoid factor. These immune complexes with hepatitis virus precipitate in the cold and are hence called cryoglobulins. They can also be entrapped in the renal glomeruli as well as in small blood vessels of the skin, nerves and other tissues, where they cause inflammation of the blood vessels (vasculitis). The most prevalent immune-complex disease is systemic lupus erythematosus (SLE), which is characterized by the formation of antibody to DNA. Every day, many millions of nuclei are extruded from erythroblasts in the bone marrow as they turn into mature red blood cells (erythrocytes). This event, among others, provides a rich source of DNA in those individuals prone to making an immune response to DNA and thereby to developing SLE.

Fig. 23.2 Three autoimmune diseases that result in damage by immune complexes.

Immune-complex disease		
Syndrome	**Autoantigen**	**Consequence**
Subacute bacterial endocarditis	Bacterial antigen	Glomerulonephritis
Mixed essential cryoglobulinemia	Rheumatoid factor IgG complexes (with or without hepatitis C antigens)	Systemic vasculitis
Systemic lupus erythematosus	DNA, histones, ribosomes, snRNP, scRNP	Glomerulonephritis, vasculitis, arthritis

Sixteen-year-old girl, butterfly rash and symmetrical morning stiffness.

The case of Nicole Chawner: too much sun at the beach.

Nicole Chawner was 16 years of age and had enjoyed good health all her life until August 1994. A few days after excessive exposure to the sun on the beach, Nicole developed a red rash on her cheeks. Her parents took her to the family doctor who recognized that the butterfly pattern of the rash on her cheeks and over the bridge of her nose was typical of systemic lupus erythematosus (SLE) (Fig. 23.3).

He referred Nicole to the Children's Hospital, where she was asked about any other problems she might have noticed, and said that when she woke up in the morning her finger joints and hips were stiff, though they got better as the day wore on. She felt equally stiff, she said, in both hands and both hips.

A blood sample was taken from Nicole to ascertain whether she had anti-nuclear antibodies (ANA). This was positive at a titer of 1:1280. In view of this result, further tests were performed for antibodies characteristically found in SLE. An elevated level of antibodies to double-stranded DNA was also found. Her serum C3 level was 73 mg dl^{-1} (normal 100–200 mg dl^{-1}). Her platelet count was normal at 225,000 μl^{-1} and her direct and indirect Coombs tests were negative, as was a test for antiphospholipid antibody. A urine sample was taken and found to be normal.

Nicole was advised to take an antimalarial agent, Plaquenil (hydroxychloroquine sulfate) and to avoid direct sunlight. (Antimalarials have a beneficial effect on SLE for unknown reasons.) She did well for a while but, after a month, the pain in her fingers and hips in the morning got worse. She developed fever of 39°C every evening accompanied by shaking chills. Enlarged lymph nodes were felt behind her ears and in the back of her neck. She lost 4.6 kg of body weight over the course of the next 2 months.

When she returned to the Children's Hospital for a check-up, it was noted that her butterfly rash had disappeared. She had diffuse swelling of the proximal joints in her fingers and toes. Blood was drawn at this time and the level of anti-DNA antibodies was found to have increased. The serum C3 level was 46 mg dl^{-1}. Nicole was advised to take 10 mg of prednisone twice a day, as well as the non-steroidal anti-inflammatory drug Naprosyn (naproxen), 250 mg twice a day. This quickly controlled her symptoms and she remained well. At her next visit to the Children's Hospital her serum C3 level was 120 mg dl^{-1}.

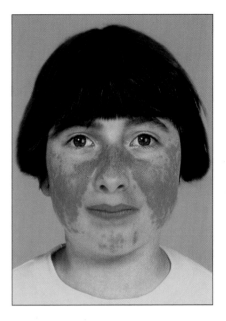

Fig. 23.3 The so-called butterfly rash typical of systemic lupus erythematosus. Photograph courtesy of M Walport.

Joint pain worse, enlarged lymph nodes. Prescribe prednisone and naproxen.

Systemic lupus erythematosus (SLE).

Systemic lupus erythematosus (SLE) is currently the most prevalent immune-complex disease in developed countries. For reasons that are not known it affects ten times as many females as males. Patients with SLE usually have antibodies to many autoantigens. The commonest autoantibody, which is found in the serum of 60% of all SLE patients, is to double-stranded DNA. Other antibodies very commonly found are to small ribonucleoproteins. Autoantibodies to blood cells, such as platelets and red blood cells, as well as to the phospholipid complex that is formed by the activation of the proteins of the clotting system (antiphospholipid antibodies), are common but less frequent. Most patients have a range of these autoantibodies.

The immune complexes in SLE are small and tend to be trapped or formed inside tissues, primarily in the kidney and, to a smaller extent, in the synovial tissues of joints. For this reason, glomerulonephritis and arthritis are two of the most frequently encountered symptoms of SLE. The immune complexes encountered in patients with SLE fix complement efficiently. Tissue injury in the kidney or joints is mediated by activation of the complement system.

The word 'lupus' is Latin for wolf and this word is applied to a common symptom of SLE, the butterfly rash on the face. In the 19th century, the severe scarring rash on the face was named lupus because it was said to resemble the bite of a wolf. At that time, it was not possible to distinguish lupus erythematosus from lupus vulgaris, a scarring rash caused by tuberculosis. For unknown reasons, the rash is evoked by exposure to the sun (ultraviolet light). There is a seasonal variation to the onset of SLE, which is greatest in the Northern Hemisphere between March and September when the greatest amount of ultraviolet light penetrates the atmosphere.

Discussion and questions.

1 Why do you think Nicole's serum C3 was measured, both on her first visit to the hospital and after therapy?

The serum level of complement proteins C3 and C4 is lowered in SLE by the large number of immune complexes that bind to them, triggering their cleavage. The depletion of these proteins is therefore proportional to the severity of the disease. Successful immunosuppressive therapy is reflected in a rise in the serum level of C3 and C4. Measurement of either C3 or C4 is sufficient: it is not necessary to measure both, and C3 is most usually measured.

2 What are the direct and indirect Coombs tests and what did they tell us in this case?

The objective of these tests was to establish whether Nicole had autoimmmune hemolytic anemia, which occurs in SLE when there are antibodies against erythrocytes. Nicole did not have hemolytic anemia (see Case 20).

3 Why was Nicole told to avoid direct exposure to sunlight?

Because ultraviolet light provokes the onset of SLE and causes relapses.

4 Repeated analysis of Nicole's urine was negative. What does this mean?

She had not developed glomerulonephritis. If she had, her urine would have contained protein and red blood cells.

5 Nicole had a serum IgG level of 1820 mg dl⁻¹. This is a substantially elevated level of IgG; this is commonly found in patients with SLE. How could you explain this? And what would you expect to find if we had a biopsy of Nicole's swollen lymph nodes?

Because of the constant stimulation of their B cells by autoantigens, patients with SLE have a greatly expanded B-cell population and consequently an increased number of plasma cells secreting immunoglobulin. A lymph node biopsy from Nicole would have exhibited follicular hyperplasia in the cortex and increased numbers of plasma cells in the medulla.

6 *The antigen in the immune complexes formed in SLE is often a complex antigen, such as part of a nucleosome or a ribonucleoprotein particle, which contains several different molecules. Patients often produce autoantibodies against each of these different components. What is the reason for the production of this variety of autoantibodies, and what type of failure in tolerance could be responsible for autoantibody production?*

In the first place, a large multimolecular complex such as a nucleosome carries many separate epitopes, each of which can stimulate antibody production by a B cell specific for that epitope. Any of these antibodies can precipitate the nucleosome particle to form an immune complex. Such potentially autoreactive B cells probably exist normally in the circulation but, provided that T-cell tolerance is intact, they are never activated because this requires T cells to be reactive against the same autoantigen. SLE is probably caused by a failure of T-cell tolerance. T cells for each of the components of the complex antigen will not be needed to induce antibodies against its individual components. As Fig. 23.4 shows, a T cell that is specific for one protein component of a nucleosome could activate B cells specific for both protein and DNA components.

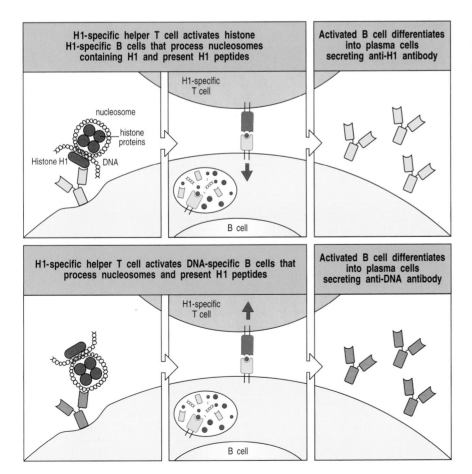

Fig. 23.4 Autoantibodies to various components of a complex antigen can be stimulated by an autoreactive helper T cell of a single specificity. In SLE, patients often produce autoantibodies to all of the components of a nucleosome, or of some other complex antigen. The most likely explanation is that all the autoreactive B cells have been activated by a single clone of autoreactive T cells specific for a peptide of one of the proteins in the complex. A B cell binding to any component of the complex through its surface immunoglobulin can internalize the complex, degrade it, and return peptides derived from the relevant protein to the cell surface bound to class II MHC molecules, where they stimulate helper T cells. These, in turn, activate the B cells. The figure illustrates this scheme for a T cell specific for the H1 histone protein of the DNA:protein complex comprising the nucleosome, and two B cells specific for the histone protein in one case, and double-stranded DNA in the other.

CASE 24 | Acute Systemic Anaphylaxis

A life-threatening immediate hypersensitivity reaction to peanuts.

Acute systemic anaphylaxis is a type I IgE-mediated hypersensitivity reaction (Fig. 24.1) in which the response is so rapid and overwhelming as to be life-threatening. As with any type I hypersensitivity reaction, the first exposure to the allergen generates allergen-specific IgE antibodies, which become bound to Fc receptors (FcεRI) on the surface of mast cells. Re-exposure to the same allergen, usually by the same route, leads to an allergic reaction due to crosslinking of the IgE antibodies by the allergen. Antibody crosslinking induces the mast cells to release a variety of chemical mediators, particularly histamine, which increases the permeability of blood vessels, and leukotrienes, which affect smooth muscle, causing bronchospasm (see Fig. 16.4).

Topics bearing on this case:
Class I hypersensitivity reactions
Allergic reactions to food
Mast-cell activation via IgE

Type I immune-mediated tissue damage	
Immune reactant	IgE antibody, T$_H$2 cells
Antigen	Soluble antigen
Effector mechanism	Mast-cell activation
Example of hyper-sensitivity reaction	Allergic rhinitis, asthma, systemic anaphylaxis

Fig. 24.1 Type I hypersensitivity reactions.

Allergens introduced systemically are most likely to cause a serious anaphylactic reaction through activation of sensitized connective tissue mast cells. The disseminated effects on the circulation and on the respiratory system are the most dangerous, while localized swelling of the throat can cause suffocation. Ingested antigens cause a variety of symptoms through their action on mucosal mast cells (Fig. 24.2).

Any allergen can provoke an anaphylactic reaction but those that most commonly cause acute systemic anaphylaxis are antibiotics, such as penicillin, and other therapeutic drugs (Fig. 24.3). These act as haptens, binding to host proteins. Proteins in food, most commonly milk, eggs, shellfish, legumes, and nuts, can also cause systemic anaphylaxis. Contact with protein antigens found in latex, a common constituent of rubber gloves, is also known to cause anaphylaxis.

Type I allergic responses are characterized by activation of CD4 helper cells (T$_H$2 cells) and IgE antibody formation. The allergen is captured by B cells through their antigen-specific surface IgM and is processed so that its peptides are presented by MHC class II molecules to T-cell receptors of antigen-specific T$_H$2 cells. The interleukins IL-4 and/or IL-13 produced by the activated T$_H$2 cells induce a switch to production of IgE, rather than IgG, by the B cell (see Fig. 3.4).

This case concerns a child who suffered from life-threatening systemic anaphylaxis caused by an allergy to peanuts.

The case of John Mason: a life-threatening immune reaction.

22-month-old child, unconscious, swollen face, difficulty breathing. Give epinephrine immediately.

John was healthy until the age of 22 months, when he developed swollen lips while eating cookies containing peanut butter. The symptoms disappeared in about an hour. A month later, while eating the same type of cookies, he started to vomit, became hoarse, had great difficulty in breathing, started to wheeze and developed a swollen face. He was taken immediately to the Emergency Room of the Children's Hospital, but on the way there he became lethargic and lost consciousness.

On arrival at hospital, his blood pressure was catastrophically low at 40/0 mmHg (normal 80/60 mmHg). Pulse was 185 beats min^{-1} (normal 80–90 beats min^{-1}), and respiratory rate 76 min^{-1} (normal 20 min^{-1}). His breathing was labored. An anaphylactic reaction was diagnosed. John was immediately given a subcutaneous injection of 0.15 ml of a 1:1000 dilution epinephrine (adrenaline). An intravenous solution of physiologic (0.15 M) saline was started at the rate of 20 ml kg^{-1} body weight h^{-1}. Also 25 mg of the anti-histamine Benadryl (diphenhydramine hydrochloride) and 25 mg of the anti-inflammatory corticosteroid Solu-Medrol (methylprednisolone) were administered intravenously. A blood sample was taken to test for histamine and the enzyme tryptase.

Within minutes of the epinephrine injection, John's hoarseness improved, the wheezing diminished and his breathing became less labored. His blood pressure rose to 50/30 mmHg, the pulse decreased to 145 beats min^{-1} and his breathing to

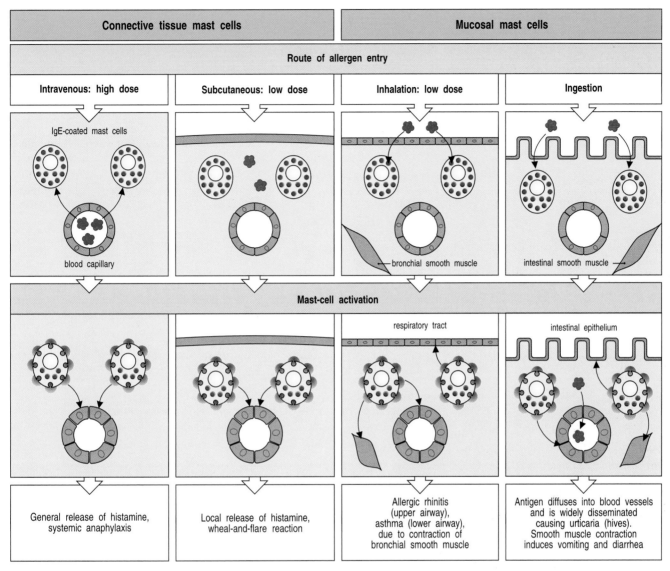

Fig. 24.2 The dose and route of allergen administration determines the type of IgE-mediated allergic reaction that results. There are two main classes of mast cell, those associated with blood vessels, called connective tissue mast cells, and those found in submucosal layers, called mucosal mast cells. In an allergic individual, all these are loaded with IgE directed against a particular antigen. Allergen in the bloodstream activates connective tissue mast cells throughout the body, resulting in the systemic release of histamine and other mediators. If very large numbers of mast cells are activated, this leads to widespread vasodilation, leading to a catastrophic loss of blood pressure, constriction of airways and swelling of the epiglottis—the anaphylactic shock syndrome. Ingested allergen penetrates across gut epithelia, causing vomiting due to smooth muscle contraction; the food allergen is also disseminated in the bloodstream, causing urticaria and other manifestations of a systemic allergic response.

61 min^{-1}. Thirty minutes later, the hoarseness and wheezing got worse again and his blood pressure dropped to 40/20 mmHg, pulse increased to 170 beats min^{-1} and respiratory rate to 70 min^{-1}.

John was given another injection of epinephrine and was made to inhale an aerosolized solution of 0.15 ml (a 5 mg ml^{-1} solution) albuterol (a β_2-adrenergic agent) in 2 ml physiologic saline. This treatment was repeated once more after 30 minutes. One hour later, the child was fully responsive, his blood pressure was

Blood pressure very low. Anaphylactic reaction.

Fig. 24.3 IgE-mediated reactions to extrinsic antigens. All IgE-mediated responses involve mast-cell degranulation, but the symptoms experienced by the patient can be very different depending on whether the allergen is injected, inhaled or eaten, and depending on the dose of the allergen.

IgE-mediated allergic reactions			
Syndrome	Common allergens	Route of entry	Response
Systemic anaphylaxis	Drugs Serum Venoms	Intravenous	Edema Vasodilation Tracheal occlusion Circulatory collapse Death
Wheal-and-flare	Insect bites Allergy testing	Subcutaneous	Local vasodilation Local edema
Allergic rhinitis (hay fever)	Pollens (ragweed, timothy, birch) Dust-mite feces	Inhaled	Edema of nasal mucosa Irritation of nasal mucosa
Bronchial asthma	Pollens Dust-mite feces	Inhaled	Bronchial constriction Increased mucus production Airway inflammation
Food allergy	Shellfish Milk Eggs Fish Wheat	Oral	Vomiting Diarrhea Pruritis (itching) Urticaria (hives)

70/50 mmHg, pulse was 116 beats min^{-1} and his respiratory rate had fallen to 46 min^{-1}. An injection of Susphrine (long-acting epinephrine) was administered, and the child was admitted to the hospital for further observation.

Treatment with Benadryl (25 mg) and methylprednisolone (1 mg kg^{-1} body weight) intravenously every 6 hours was continued for 24 hours, by which time the facial swelling had subsided, and John's blood pressure, respiratory rate and pulse were normal. He had stopped wheezing and when the doctor listened to his chest with a stethoscope it was clear.

John was taken off all medication and observed for another 12 hours. He remained well and was discharged home. His parents were instructed to avoid giving him foods containing peanuts in any form, and were asked to bring him to the Allergy Clinic for further tests in a few days' time.

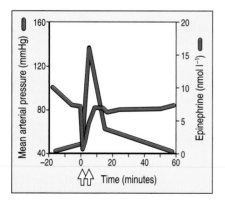

Fig. 24.4 Mean arterial pressure and epinephrine levels in a representative patient with insect-sting anaphylactic shock. Time O indicates the onset of the anaphylactic reaction as reported by the patient. The arrows indicate administration of anti-histamines and epinephrine.

Acute systemic anaphylaxis.

Anaphylaxis presents a medical emergency and is the most urgent of clinical immunologic events; it requires immediate therapy. It results from the generation and release of a variety of potent biologically active mediators and their concerted effects on a number of target organs. John showed classic rapid-onset symptoms of anaphylaxis, starting with vomiting and swelling of the face and throat, and constriction of the bronchial smooth muscle, which led to his difficulty in breathing. This was soon followed by a catastrophic loss of blood pressure, due to leakage of fluid from the blood vessels. Anaphylaxis can also cause urticaria (hives), heart arrhythmias, and myocardial ischemia, and gastrointestinal symptoms such as nausea, pain, and diarrhea. All these signs and symptoms can occur singly or in combination (Fig. 24.4).

Fatal allergic reactions to the venoms in bee and wasp stings have been recognized for at least 4500 years and account today for approximately 40 deaths each year in the USA. In 1902, Portier and Richet reported that a second injection of a protein from a sea anemone caused a fatal systemic reaction in dogs that had been injected previously with this protein. Since this form of immunity was fatal rather than protective, it was termed anaphylaxis to distinguish it from the prophylaxis (protection) generated by immunization.

Anaphylaxis requires a latent period for sensitization after the first introduction of antigen followed by re-exposure to the sensitizing agent, which can be any foreign protein or a hapten. In the early part of the twentieth century the most frequent cause of systemic anaphylaxis was horse serum, which was used as a source of antibodies to treat infectious diseases. The widespread use of penicillin now makes it a major culprit in anaphylactic death in the USA, with a current estimated mortality of around 100 cases each year. Newly introduced chemicals and drugs add continually to the list of agents capable of causing anaphylaxis.

The rate of fatal anaphylaxis from any cause is estimated at 0.4 cases per million individuals per year. The risk of non-fatal anaphylaxis is difficult to assess and has been estimated to range from 0.1% to 1% for people to whom penicillin is administered or who are stung by bees or wasps. Although, in John's case, the reaction was brought on by eating a food, an antigen administered by subcutaneous, intramuscular or intravenous injection is more likely to induce a clinical anaphylactic reaction than one that enters by the oral or respiratory route. Atopic individuals (see Case 16) are no more susceptible to anaphylaxis than non-atopic individuals.

Discussion and questions.

---

1   *Anaphylaxis results in the release of a variety of chemical mediators from mast cells, such as histamine and leukotrienes. Angioedema (localized swelling caused by an increase in vascular permeability and leakage of fluid into tissues) is one of the symptoms of anaphylaxis. With the above in mind, why did John get hoarse and why did he wheeze?*

Johns hoarseness resulted from angioedema of the vocal cords. His wheezing was due to forced expiration of air through bronchi that had become constricted. In this case, constriction resulted from the release by activated mast cells of histamine and leukotrienes that caused the smooth muscles of the bronchial tubes to constrict.

2   *When his parents brought John back to the Allergy Clinic, a nurse performed several skin tests by pricking the epidermis of his forearm with a shallow plastic needle containing peanut antigens. John was also tested in a similar fashion with antigens from nuts as well as from eggs, milk, corn, and wheat. Within 5 minutes John developed a wheal, 10 x 12 mm in size, surrounded by a red flare, 25 x 30 mm (see Fig. 16.7), at the site of application of the peanut antigen. No reactions were noted*

to the other antigens. A radioallergosorbent test (RAST) was performed on a blood sample to examine for the presence of IgE antibodies to peanut antigens. It was positive. What would you advise John's parents to do?

John's parents were instructed to avoid feeding him any food containing peanuts and to read labels of packaged foods scrupulously to avoid anything containing peanuts. They were advised to inquire in restaurants about food containing peanuts. Because green peas, also a legume, contain an antigen that cross-reacts with peanuts and might also incite an anaphylactic reaction, peas were withdrawn from John's diet. A Medi-Alert bracelet, indicating his anaphylactic reaction to peanuts, was ordered for John. The parents were also given an Epi-Pen syringe pre-filled with epinephrine to keep at home or while travelling, in case John developed another anaphylactic reaction.

3 *Why was John treated first with epinephrine in the emergency room?*

Epinephrine acts at β_2-adrenergic receptors in smooth muscle surrounding blood vessels and bronchi. It has opposing effects on the two types of muscle. It contracts that surrounding the small blood vessels, thereby constricting them, stopping vascular leakage and raising the blood pressure. It relaxes that of the bronchi, making breathing easier.

4 *What other drug in the epinephrine family was John given?*

Albuterol, which is also a β_2-adrenergic agent. It can be inhaled to alleviate bronchial constriction more effectively.

5 *Why was John given a blood test for histamine and the enzyme tryptase?*

Histamine and tryptase are released by activated mast cells; high levels in the blood indicate the massive release from the mast cells that occurs during an anaphylactic reaction.

6 *Why was the skin testing for peanuts not done in the hospital immediately after John had recovered, rather than at a later visit?*

Immediately after a systemic anaphylactic reaction the patient is unresponsive in a skin test owing to the massive depletion of mast cell granules and failure of the blood vessels to respond to mediators. This is called tachyphylaxis and lasts for 72–96 hours after the anaphylactic reaction. For this reason, John had to come back to the Allergy Clinic a few days later for his tests.

CASE 25 | Drug-Induced Serum Sickness

An adverse immune reaction to an antibiotic.

The intravenous administration of a large dose of antigen can evoke in some individuals a type III hypersensitivity reaction or immune-complex disease (Fig. 25.1). Antigen administration produces a rapid IgG response and the formation of antigen:antibody complexes (immune complexes) that can activate complement.

Owing to the large amount of antigen present and the rapid IgG response, small immune complexes begin to be formed in conditions of antigen excess (Fig. 25.2). Unlike the large immune complexes that are formed in conditions of antibody excess, which are rapidly ingested by phagocytic cells and cleared from the system, the smaller immune complexes are taken up by endothelial cells in various parts of the body and become deposited in tissues. Local activation of the complement system by these immune complexes provokes localized inflammatory responses.

The experimental model for immune-complex disease is the Arthus reaction, in which the subcutaneous injection of large doses of antigen evokes a brisk IgG response. The activation of complement by the IgG:antigen complexes generates the complement component C3a, a potent stimulator of histamine release from mast cells, and C5a, one of the most active chemokines produced by the body. The local endothelial cells are activated by the inter-actions in blood vessels between the immune complexes, complement and circulating leukocytes and platelets. They upregulate their expression of adhesion molecules such as selectins and integrins, which facilitates the emigration of white blood cells from the bloodstream and the initiation of a local inflammatory reaction (Fig. 25.3). Platelets also accumulate at the site, causing blood clotting; the small blood vessels become plugged with clots and burst, producing hemorrhage in the skin (Fig. 25.4).

When an antigen is injected intravenously, the immune complexes formed can be deposited at a wide range of sites. When deposited in synovial tissue, the resulting inflammation of the joints produces arthritis; in the kidney glomeruli they cause glomerulonephritis; and in the endothelium of the blood vessels of the skin and other organs they provoke vasculitis (Fig. 25.5).

Topics bearing on this case:

Hypersensitivity reactions

Immune-complex formation

Properties of IgG antibodies

Activation of complement by antigen:antibody complexes

Inflammatory reactions

Type III immune-mediated tissue damage	
Immune reactant	IgG antibody
Antigen	Soluble antigen
Effector mechanism	Complement Phagocytes
	immune complex + complement
Example of hyper-sensitivity reaction	Serum sickness, systemic lupus erythematosus

Fig. 25.1 Type III hypersensitivity reactions. These can be caused by large intravenous doses of soluble antigens (serum sickness) or by an autoimmune reaction against some types of self antigen. The IgG antibodies produced form small immune complexes with the antigen in excess. The tissue damage involved is caused by complement activation and the subsequent inflammatory responses, which are triggered by immune complexes deposited in tissues.

In the early years of the twentieth century, the most common cause of immune-complex disease was the administration of horse serum, which was used as a source of antibodies to treat infectious diseases, and so this type of hypersensitivity reaction to large doses of intravenous antigen is still known as serum sickness.

This case describes a 12-year-old boy who received massive intravenous injections of penicillin and ampicillin (one of its analogs) to treat pneumonia. He developed a serum sickness reaction to the antibiotics.

The case of Gregory Barnes: serum sickness precipitated by penicillin.

When Gregory was brought to the Children's Hospital Emergency Room, his parents told the physicians that for 2 days he had had high fever (>39.5°C), a cough and shortness of breath. Before then he had enjoyed excellent health. On physical examination he was pale, looked dehydrated and was breathing rapidly with flaring nostrils. His respiratory rate was 62 min^{-1} (normal 20 min^{-1}), pulse was 120 beats min^{-1} (normal 60–80 beats min^{-1}), blood pressure was 90/60 mmHg (normal). When his chest was examined with a stethoscope the Emergency Room doctors heard crackles (bubbly sounds) over the lower left lobe of his lungs. A chest radiograph revealed an opaque area over the entire lower lobe of the left lung. A diagnosis of lobar pneumonia was made.

A white blood count revealed 19,000 cells μl^{-1} (normal 4000–7000 cells μl^{-1}) with an increase in the percentage of neutrophils to 87% of total white blood cells (normal 60%) and the abnormal presence of immature forms of neutrophils. A Gram stain of Gregory's sputum revealed Gram-positive cocci. Sputum and blood cultures grew *Streptococcus pneumoniae* (the pneumococcus).

Gregory was admitted to the hospital and treated with intravenous ampicillin at a dose of 1 g every 6 hours. Gregory gave no history of allergy to penicillin so ampicillin was used to cover both Gram-positive and Gram-negative bacteria. On the fourth day of treatment, he felt remarkably better, his respiratory rate had decreased to 40 min^{-1}, and his temperature to 37.5°C. His white cell count had

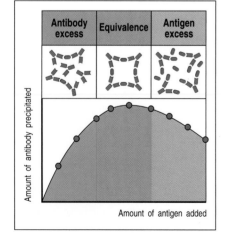

Fig. 25.2 Antibody can precipitate soluble antigen in the form of immune complexes. *In vitro*, the precipitation of immune complexes formed by antibody crosslinking the antigen molecules can be measured and used to define zones of antibody excess, equivalence, and antigen excess. In the zone of antigen excess, some immune complexes are too small to precipitate. When this happens *in vivo* such soluble immune complexes can produce pathological damage to blood vessels.

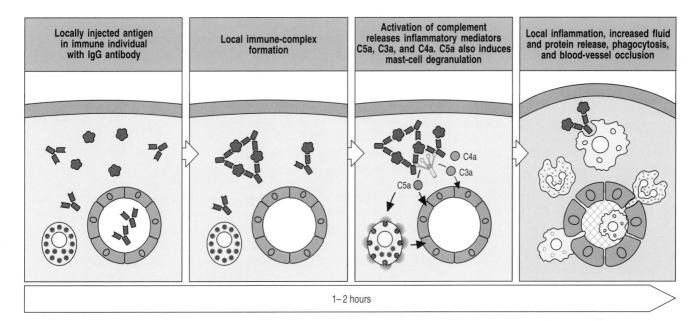

| Locally injected antigen in immune individual with IgG antibody | Local immune-complex formation | Activation of complement releases inflammatory mediators C5a, C3a, and C4a. C5a also induces mast-cell degranulation | Local inflammation, increased fluid and protein release, phagocytosis, and blood-vessel occlusion |

1–2 hours

Fig. 25.3 The deposition of immune complexes in local tissues causes a local inflammatory response known as an Arthus reaction. In individuals who have already made IgG antibodies to an allergen, the same allergen injected into the skin forms immune complexes with IgG antibody that has diffused out of the capillaries. Since the dose of antigen is low, the immune complexes are only formed close to the site of injection, where they activate complement, releasing inflammatory mediators such as C5a, which in turn can activate mast cells to release inflammatory mediators. As a result inflammatory cells invade the site and blood vessel permeability and blood flow are increased. Platelets also accumulate at the site, ultimately leading to occlusion of the small blood vessels, hemorrhage and the appearance of purpura.

decreased to 9000 μl⁻¹. Because the *S. pneumoniae* grown from his sputum and blood was sensitive to penicillin, the ampicillin was replaced by penicillin. On his ninth day in hospital, Gregory had no fever, his white cell count was 7000 μl⁻¹ and his chest radiograph had improved. Plans for discharge from hospital were made for the following day.

The next morning, Gregory had puffy eyes, and welts resembling large hives on his abdomen. He was given the anti-histamine Benadryl (diphenhydramine hydrochloride) orally and penicillin was discontinued. Two hours later he developed a tight feeling in the throat, a swollen face, and widespread urticaria (hives). With a stethoscope, wheezing could be heard all over his lungs. The wheezing responded to inhalation of the β₂-adrenergic agent, albuterol. That evening Gregory developed a fever (a temperature of 39°C), swollen and painful ankles and his urticarial rash became more generalized. He appeared once again acutely ill.

Allergy developing discontinue penicillin immediately.

The rash spread over his trunk, back, neck and face, and in places had become confluent (Fig. 25.6). Gregory also had reddened eyes owing to inflamed conjunctivae, and had swelling around the mouth. The anterior cervical, axillary and inguinal lymph nodes on both sides were enlarged, measuring 2 × 1 cm in diameter. The spleen was also enlarged, with its tip palpable 3 cm below the rib margin. Ankles and knee joints were swollen and tender to palpation, and were too painful to move very far. The child was alert and his neurologic examination was normal.

Laboratory analysis of a blood sample revealed a raised white blood cell count (19,800 μl⁻¹) in which the predominant cells were lymphocytes (72% compared with the normal 30%). Plasma cells were detected in a blood smear, although plasma cells are normally not present in blood. The erythrocyte sedimentation rate, an

Fig. 25.4 Hemorrhaging of the skin in the course of a serum sickness reaction.

Route	Resulting disease	Site of immune complex deposition
Intravenous (high dose)	Vasculitis	Blood vessel walls
	Nephritis	Renal glomeruli
	Arthritis	Joint spaces
Subcutaneous	Arthus reaction	Perivascular area
Inhaled	Farmer's lung	Alveolar/capillary interface

Fig. 25.5 The dose and route of antigen delivery determines the pathology observed in type III allergic reactions.

Serum complement components low. Serum sickness.

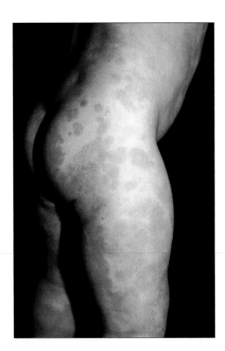

Fig. 25.6 Urticarial rash as a consequence of a serum sickness reaction.

indicator of the presence of acute-phase reactants in the blood, was elevated at 30 mm h^{-1} (normal <20 mm h^{-1}). His total serum complement level and his serum C1q and C3 levels were decreased.

A presumptive diagnosis of serum sickness was made and Gregory was given Benadryl and Naprosyn (naproxen), a non-steroidal anti-inflammatory agent. On the following day, the rash and joint swellings were worse and the child complained of abdominal pain. There were also purpuric lesions, caused by hemorrhaging of small blood vessels under the skin, on his feet and around his ankles. There was no blood in his stool.

Later in the day, Gregory became agitated, and had periods of disorientation when his speech was unintelligible and he could not recognize his parents. A CT scan of his brain proved negative, as did an examination of his cerebrospinal fluid for the presence of inflammatory cells, increased protein concentration and decreased sugar concentration, all of which are indicators of infection and inflammation. However, his electroencephalogram was abnormal, with a pattern that suggested diminished circulation in the posterior part of the brain.

His white blood count rose to 23,700 cells ml^{-1} and his erythrocyte sedimentation rate to 54 mm h^{-1}. Red cells and protein were now present in the urine. A skin biopsy from a purpuric area on his foot showed moderate edema (swelling) around the capillaries and in the dermis, as well as perivascular infiltrates of lymphocytes in the deeper dermis. Immunofluorescence microscopy of the biopsy tissue with the appropriate antibodies revealed deposition of IgG and C3 in the perivascular areas.

Gregory was started on the anti-inflammatory corticosteroid prednisone and all his symptoms improved progressively; the joint swelling and splenomegaly resolved over the next few days. He was soon able to walk and was discharged 7 days after the onset of his serum sickness on a slowly decreasing course of prednisone and Benadryl. On follow-up examination 2 weeks later, Gregory had no IgE antibodies to penicillin or ampicillin, as detected by both immediate hypersensitivity skin tests and by an *in vitro* radioallergosorbent test (RAST). His parents were instructed that Gregory should never be given any penicillin, penicillin derivatives or cephalosporins.

Serum sickness.

The classic symptoms of serum sickness that Gregory showed were first described in great detail by Clemens von Pirquet and Bela Schick in a famous monograph entitled *Die Serumkrankheit* (Serum Sickness), published in 1905. Schick subsequently translated this monograph into English and it was re-issued by Williams and Wilkins in 1951. It is fascinating to read this short work in the light of current knowledge. In the 1890s it had become common practice to treat diphtheria with horse serum containing antibodies taken from horses that had been immunized with diphtheria toxin. Immune horse serum was also used to treat scarlet fever, which was then a life-threatening illness. Von Pirquet and Schick made systematic observations on tens of children who developed the symptoms and signs of serum sickness at the St Anna's Children's Hospital in Vienna and described the classic symptoms of the disease. They correctly surmised that serum sickness was due to an immunologic reaction to horse serum proteins in their patients (Fig. 25.7).

Experimental models of serum sickness were developed in the 1950s by Hawn, Janeway and Dixon, who injected rabbits with large amounts of bovine serum albumin or bovine gamma globulin. They noted that the rabbits developed glomerulonephritis just at the time when antibody to the foreign protein first appeared in the rabbit serum, accompanied by a profound and transitory fall in the serum complement level. By this time immunochemistry had advanced to the point where it was possible to show that the disease was caused by the formation and deposition of small immune complexes.

Although horse serum is no longer used in therapy, other foreign proteins are still administered to patients. Antitoxins to snake venom are produced in various animal species and mouse monoclonal antibodies are used in clinical practice. However, the commonest causes of serum sickness today are antibiotics, particularly penicillin and its derivatives, which act as haptens. These drugs bind to host proteins that serve as carriers and thus can elicit a rapid and strong IgG antibody response.

Serum sickness, although very unpleasant, is a self-limited disease that terminates as the immune response of the host moves into the zone of antibody excess. It can prove fatal if it provokes kidney shutdown or bleeding in a critical area such as the brain. Its course can be ameliorated by anti-inflammatory drugs such as prednisone and anti-histamines. It is also unlike the other types of hypersensitivity in that a reaction can appear on first encounter with the antigen, if it is long-lived and given in a sufficiently large dose. This seems to have been the case for Gregory.

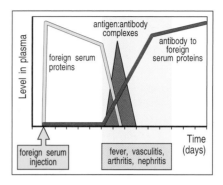

Fig. 25.7 Serum sickness is the classic example of a transient syndrome mediated by immune complexes. An injection of large amounts of foreign proteins, in this case derived from horse serum, leads to an antibody response. These antibodies form immune complexes with the circulating foreign proteins. These complexes activate complement and phagocytes, inducing fever, and are deposited in small blood vessels, inducing the symptoms of vasculitis, glomerulo-nephritis, and arthritis. All these effects are transient and resolve when the foreign protein is cleared from the system.

Discussion and questions.

1 *Hives (urticaria) and edema about the mouth and eyelids were the first symptoms of serum sickness developed by Gregory. What caused these early symptoms?*

The activation of complement-generated C3a, which releases histamine from mast cells and causes hives. The swelling around the mouth and eyelids is a form of angioedema. There is a more complete discussion of the role of the complement and the kinin systems in the pathogenesis of angioedema in Case 9.

2 *At one point Gregory became confused and disoriented and did not recognize his parents. His cerebrospinal fluid was normal and a CT scan of his brain was normal. However, an electroencephalogram displayed an abnormal pattern of brain waves. What produced these clinical and laboratory findings?*

Gregory almost certainly had developed vasculitis in the small blood vessels of his brain and this compromised oxygen delivery to his brain.

3 *What other manifestations of vasculitis were noted in Gregory?*

He had red cells and albumin in his urine, which indicated an inflammation of the small blood vessels in his kidney glomeruli. He also developed purpura in his feet and ankles. Purpura (which is the Latin word for purple) indicates hemorrhage from small blood vessels in the skin that are inflamed and have become plugged with clots. A skin biopsy of one lesion showed the deposition of IgG and C3 around the small blood vessels, suggesting that an immune reaction was taking place.

4 *Gregory had enlarged lymph nodes everywhere and his spleen was also enlarged. If you had a biopsy of a node what would you expect to see?*

Massive follicular hyperplasia, polyclonal B-cell activation and many mature plasma cells in the medulla. The massive B-cell activation in the lymph nodes leads to an overflow of plasma cells from the medulla of the nodes into the efferent lymph. It is otherwise very, very rare to find plasma cells in the blood, as were found in Gregory's blood. They find their way to the bloodstream via the thoracic duct. The enlargement of the spleen was almost certainly due to hyperplasia of the white pulp. Some plasma cells probably enter the blood from the hyperplastic follicles in the spleen.

5 *Gregory had a brisk 'acute-phase response'. What is it and what causes it?*

The acute-phase reaction is caused by interleukin-1 and to a greater extent by interleukin-6, which are released from monocytes that have been activated by the uptake of immune complexes. The acute-phase response consists of dramatic changes in protein synthesis by the liver. The synthesis of albumin drops sharply, as does the synthesis of transferrin. The synthesis of fibrinogen, C-reactive protein, amyloid A and several glycoproteins is rapidly upregulated. The precise advantage to the host of the acute-phase reaction is not well understood, but it is presumably a part of innate immunity, which aids host resistance to pathogens before the adaptive immune system becomes engaged.

6 *Penicillin can cause more than one type of hypersensitivity reaction. What laboratory test gave the best evidence that Gregory was suffering from a disease caused by immune complexes?*

His serum C1q level was decreased. This almost always indicates complement consumption by immune complexes via the classical pathway. (In hereditary angioneurotic edema (see Case 9) the C1q level is normal; in this disease the activation of complement is not caused by immune complexes.) The level of C3 in Gregory's serum was also lowered: a further indication of complement consumption (see Case 23).

7 *When Gregory returned for a follow-up clinic visit, a skin test for immediate hypersensitivity was performed by intradermal injection of penicillin. He did not respond. Does this mean that an incorrect diagnosis was made and that he did not have serum sickness due to penicillin?*

No! The skin test is positive when there are IgE antibodies bound to the mast cells in the skin. Gregory did not have IgE antibodies to penicillin, as confirmed by the negative RAST test. Serum sickness is caused by complement-fixing IgG antibodies.

CASE 26 | Multiple Myeloma

A malignancy of terminally differentiated B lymphocytes.

Malignant tumors result from the outgrowth of a single transformed cell. Malignant tumors that result from the clonal outgrowth of B lymphocytes can occur at all stages of B-cell development (Fig. 26.1). These malignant B cells express clonotypic immunoglobulin molecules derived from the same immunoglobulin gene rearrangement, either on their surface or as secreted monoclonal antibody.

Malignancies of plasma cells cause a disease called multiple myeloma. It is a disease of bone because these plasma-cell tumors arise in the bone marrow. As the tumor masses expand, they cause local erosions of the bone, and the appearance on radiographs of multiple bone lesions (Fig. 26.2).

Fig. 26.1 B-cell tumors represent clonal outgrowths of B cells at various stages of development. Each type of tumor cell has a normal B-cell equivalent, homes to similar sites, and has behavior similar to that cell. Thus, myeloma cells look much like the plasma cells from which they derive, they secrete immunoglobulin, and they are found predominantly in the bone marrow. Many lymphomas and myelomas may go through a preliminary less aggressive lymphoproliferative phase, and some mild lymphoproliferations appear to be benign.

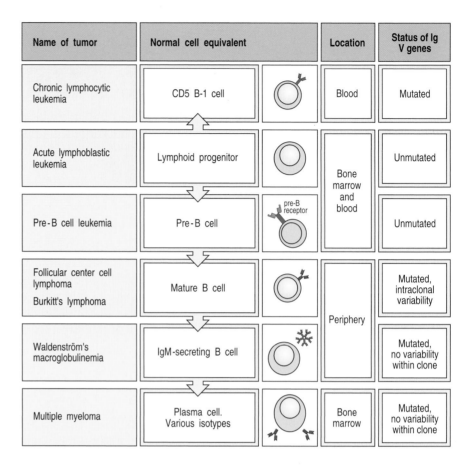

Name of tumor	Normal cell equivalent		Location	Status of Ig V genes
Chronic lymphocytic leukemia	CD5 B-1 cell		Blood	Mutated
Acute lymphoblastic leukemia	Lymphoid progenitor		Bone marrow and blood	Unmutated
Pre-B cell leukemia	Pre-B cell	pre-B receptor		Unmutated
Follicular center cell lymphoma / Burkitt's lymphoma	Mature B cell		Periphery	Mutated, intraclonal variability
Waldenström's macroglobulinemia	IgM-secreting B cell			Mutated, no variability within clone
Multiple myeloma	Plasma cell. Various isotypes		Bone marrow	Mutated, no variability within clone

These myelomas secrete a staggering amount of monoclonal immunoglobulin, which may be of the IgG or IgA, or very rarely IgD or IgE, isotype, bearing either kappa (κ) or lambda (λ) light chains (Fig. 26.3). The malignant plasma cells asynchronously synthesize more light chains than heavy chains, so that immmunoglobulin light chains are excreted in the urine in excessive amounts. In 1846 Dr Charles McIntyre, a physician practicing medicine in London, made a house call on a greengrocer residing in Devonshire Street. Seeing this unfortunate man wasting away with fragile bone disease, it occurred to Dr McIntyre that he might be losing excessive amounts of protein

Fig. 26.2 Radiographs of the skull and a long bone in a patient with multiple myeloma. Note the 'punched out' lesions in the bones, where the accumulation of malignant plasma cells has eroded the normal calcification. Courtesy of L Shulman.

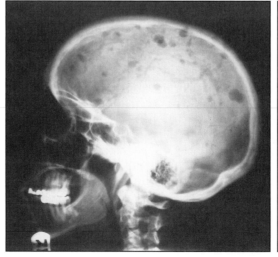
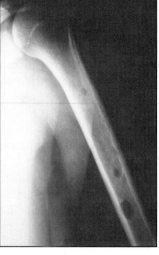

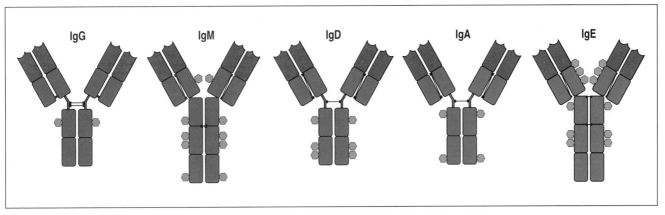

Fig. 26.3 The structural organization of the main human immunoglobulin isotype monomers. The choice of constant-region gene determines the class or isotype of the immunoglobulin made. Both IgM and IgE lack a hinge region but each contains an extra heavy-chain C domain. The isotypes also differ in the distribution of N-linked carbohydrate groups, as shown in turquoise, and in the distribution of disulfide bonds.

('animal matter') in his urine. He took a urine specimen back to his consulting rooms and found that, upon heating, a precipitate formed in the urine between 45 and 60°C, and re-dissolved upon further heating of the urine to boiling point. On the following day, he sent a specimen of this urine to Dr Henry Bence-Jones, Professor of Clinical Chemistry at Guy's Hospital, with a complete description of the bizarre characteristics of the abnormal protein that precipitated, and a question: 'What is this?'. The protein was thenceforth known as a Bence-Jones protein. It took over 100 years to answer the question but, eventually, Bence-Jones proteins were found to be immunoglobulin light chains.

The case of Isabel Archer: the consequences of unrestrained growth of an antibody-producing B-cell clone.

Isabel Archer was a 55-year-old housewife in 1989, when she began to experience excessive fatigue. She had been in good health her entire life. Her 57-year-old husband, a successful lawyer, was also in good health, as were her three sons, all in their 20s. At the time of a routine annual check-up at her physician she reported to him how easily she became fatigued. He found no abnormalities on physical examination.

A blood sample revealed that she had mild anemia; her red blood cell count was $3.5 \times 10^6\ \mu l^{-1}$ (normal $4.2-5.0 \times 10^6\ \mu l^{-1}$). Her white blood cell count was $3600\ \mu l^{-1}$ (normal $5000\ \mu l^{-1}$). The sedimentation rate of her red blood cells was $32\ mm\ h^{-1}$ (normal $<20\ mm\ h^{-1}$). Unclotted whole blood from Mrs Archer was put in a narrow-bore tube to determine how far the red blood cells would sediment in 1 hour. Sedimentation of the red blood cells is caused by rouleaux formation, in which red blood cells stack on one another, and is hastened when the fibrinogen or IgG

Mrs Archer, 55 years, previously very healthy but now easily fatigued.

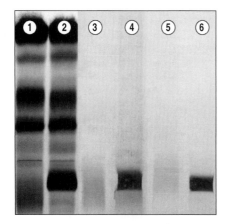

Fig. 26.4 Electrophoresis indicates whether serum immunoglobulins have monoclonal components. An electrophoresic pattern of normal serum run on an agarose gel (lane 1) is shown next to the pattern obtained with a serum sample from Mrs Archer (lane 2). The heterogeneous immunoglobulins from normal serum stained as a smear, whereas the monoclonal component of Mrs Archer's serum ran as a tight protein band. The electrophoresis was performed again with normal serum (lanes 3 + 5) and Mrs Archer's serum (lanes 4 + 6) and this reacted with an antibody to γ chains (lanes 3 + 4) and antibody to κ chains (lanes 5 + 6). The agarose gel was washed to remove all proteins except for antigen:antibody complexes. This shows that Mrs Archer's myeloma protein was IgGκ.

Elevated sedimentation rate, elevated IgG. IgG levels continue to rise.

content of the blood plasma is elevated. This elevated sedimentation rate prompted the measurement of her serum immunoglobulins. The concentration of IgG was found to be 3790 mg dl^{-1} (normal 600–1500 mg dl^{-1}), that of IgA 14 mg dl^{-1} (normal 150–250 mg dl^{-1}) and that of IgM 53 mg dl^{-1} (normal 75–150 mg dl^{-1}). Electrophoresis of her serum revealed the presence of a monoclonal protein, which on further analysis was found to be IgG with kappa light chains (Fig. 26.4). Radiographs of all of her bones did not show any abnormality. No treatment was advised.

Mrs Archer returned for regular visits to her physician and on each occasion he measured her serum IgG level and noticed that it was gradually increasing. In April 1991 her serum IgG was 4520 mg dl^{-1}, and in January 1992 it was 5100 mg dl^{-1}. By November 1992, her anemia had worsened and her red blood cell count had fallen to 3.0×10^6 μl^{-1}. At the same time her white blood count had fallen to 2600 μl^{-1}.

Plasmacytoma in second thoracic vertebral body.

In December 1992, Mrs Archer experienced the sudden onset of upper back pain. She was referred to a radiologist who performed a radiograph of the thoracic spine followed by a magnetic resonance imaging (MRI) scan. He reported to the internist that he found destruction of the second thoracic vertebral body with extrusion of a plasmacytoma (a tumor of plasma cells) from the affected vertebral body compressing the spinal cord. Mrs Archer was treated with the corticosteroid, decadron, and irradiation to her spine. Her symptoms improved. However, her serum IgG level reached 6312 mg dl^{-1} and she required blood transfusions because of her worsening anemia. She was treated with melphalan and prednisone.

In April 1993, further chemotherapy was given because of the persisting elevation of her serum IgG. She was treated for 9 months with vincristine, adriamycin and decadron and her serum IgG fell from 6785 mg dl^{-1} to 5308 mg dl^{-1}. When her serum IgG subsequently rose to 8200 mg dl^{-1} she was treated with a course of cyclophosphamide, etoposide and decadron, which reduced her serum IgG level to 6000 mg dl^{-1}.

IgG level keeps rising.

In February 1995, Mrs Archer developed high fever and chest pain. On chest radiograph she was found to have pneumonia of the lower lobe of the left lung. She was treated successfully with antibiotics. She again experienced high fever, shaking chills and chest pains in May 1995.

Because she was hypotensive (low blood pressure) she was admitted to the intensive care unit and given antibiotics intravenously and cardiac pressors to raise her blood pressure. *Streptococcus pyogenes* was cultured from her sputum and blood. She recovered from this episode in the hospital and remains fully active. She requires occasional blood transfusions for her anemia and complains at times of bone pain. Her serum IgG is stable at 6200 mg dl^{-1}.

Multiple myeloma.

Isabel Archer presents us with many, if not most, of the typical features of multiple myeloma, a malignant tumor of plasma cells. A single plasma cell has undergone malignant transformation; its progeny have disseminated to many sites in the bone marrow and are producing prodigious quantities of a monoclonal immunoglobulin.

Multiple myeloma is a very malignant disease that is resistant to most cancer chemotherapy. Methylphenylalanine mustard (melphalan), which Mrs Archer received, is one of the few chemotherapeutic agents that has been effective in the treatment of this disease. Although Mrs Archer was in relatively good health as our case history ended, her outlook for survival is very poor. Recently, bone marrow transplants have been used to cure patients with multiple myeloma.

Myeloma proteins have played an important part in the history of immunology. Subclasses of IgG were first recognized, for example, when a rabbit was immunized with a single human myeloma protein and found to react with 80% of myeloma proteins but not with the other 20%. This led to the conclusion that the 80% that did react belonged to an IgG subclass (IgG1) capable of generating subclass-specific antibodies in rabbits. Four subclasses of IgG were distinguished by immunizing rabbits with single myeloma proteins and testing the antibodies generated for cross-reactivity to other myeloma proteins. Korngold and Lipari had already used this approach to classify Bence-Jones proteins into two groups of proteins, subsequently called kappa and lambda light chains. Later on, a myeloma protein that was available in abundant quantity as a homogenous protein became the first immunoglobulin molecule for which a complete amino acid sequence was obtained.

Discussion and questions.

1 The serum IgG from Mrs Archer was assumed to be monoclonal because it migrated as a tight band on electrophoresis in an agarose gel, and because it reacted with antibodies to kappa but not to lambda chains. What other evidence could be brought to bear to prove the monoclonality of this IgG?

The IgG could also be shown to belong to a single subclass of IgG, that is IgG1, IgG2, IgG3, or IgG4. Furthermore, it would be possible to show that a single variable-region gene was rearranged to form this IgG.

2 Mrs Archer became anemic (low red blood cell count) and neutropenic (low white blood cell count). What was the cause of this?

The proliferation of malignant plasma cells in the bone marrow crowded out the red blood cell and white blood cell precursors. This creates a limitation on space in the bone marrow.

3 As her disease progressed, Isabel Archer became susceptible to pyogenic infections; for example, she had pneumonia twice in a short period. What is the basis of her susceptibilty to these infections?

Although her serum IgG concentration is quite elevated, almost all the IgG is secreted by the myeloma cells and is monoclonal. In fact, she has very little normal polyclonal IgG and has been effectively rendered agammaglobulinemic by her disease. In addition, her white blood cell count is decreased and she has too few neutrophils ($<1000 \ \mu l^{-1}$) to ingest bacteria in the bloodstream and lungs effectively.

4   You might conclude that it would be useful to administer gamma globulin intravenously to Mrs Archer to protect her from more pyogenic infections. Why would this treatment be less successful than in the case of X-linked agammaglobulinemia?

We learned in the case of X-linked agammaglobulinemia that the turnover or fractional catabolic rate of IgG is concentration-dependent. In the young man with X-linked agammaglobulinemia, the administered IgG lasted longer than it would in a normal person. In Mrs Archer, the very elevated level of serum IgG would result in rapid catabolism of the IgG administered.

5   A monoclonal immunoglobulin in the serum is called an M-component ('M' for myeloma). Is the presence of an M-component in serum diagnostic of multiple myeloma?

No. M-components appear in the blood as people age. About 10% of healthy individuals in the ninth decade of life have M-components. This is called benign monoclonal gammopathy. Without bone lesions and the presence of many malignant plasma cells in the bone marrow, the diagnosis of multiple myeloma cannot be made. Some people have IgM M-components in their blood. This is due to another malignancy of plasma cells called Waldenstrom's macroglobulinemia, which differs in many ways from multiple myeloma and is a more benign disease.

6   Very rarely an individual with multiple myeloma has two M-components in the blood. Although these two M-components derive from different constant-region genes, their antigen-binding regions are both encoded by the same variable-region gene. Can you hypothesize how this happens?

The best explanation is that the malignant transformation of a plasma cell occurred as class switching was taking place. The progeny of this cell can make two immunoglobulin classes, for example IgM and IgG. It is not known whether some cells make IgM and some IgG, or all of the cells make both, although it seems most likely that each cell makes only one isotype.

CASE 27 | T-Cell Lymphoma

A malignancy of functional T lymphocytes.

T-cell lymphomas arise when mature T cells or their precursors in the thymus gland undergo malignant transformation and clonal expansion. These tumors seem to represent arrested stages of normal differentiation, and express molecules on their surface that are normally present on T cells at various stages of maturation (Fig. 27.1). Occasionally, T-cell lymphomas even mimic the immune effector functions of their non-transformed normal counterparts.

T cells develop in the thymus gland, where precursors arriving from the bone marrow enter a phase of intense proliferation. As thymocytes proliferate and mature into T cells, they pass through a series of distinct phases (Fig. 27.2; see also Fig. 5.3). These are marked by changes in the status of T-cell receptor genes and the expression of T-cell receptor proteins on the T-cell surface, and by changes in the co-receptor molecules CD4 and CD8. Particular combinations of these cell-surface molecules can thus be used as markers for T cells at different stages of differentiation.

When progenitor T cells first enter the thymus from the bone marrow, their receptor genes are unrearranged and they lack most of the surface molecules characteristic of mature T cells. Interactions with the thymic stroma trigger rapid cell proliferation and expression of the first T-cell-specific surface molecule, CD2. At the end of this phase, the immature thymocytes bear

Topics bearing on this case:
Development of T cells
T-cell tumors
Role of cytokines in isotype switching and leukocyte production

Fig. 27.1 T-cell tumors represent monoclonal outgrowths of normal cell populations. Each distinct T-cell tumor has a normal equivalent, as also seen with some B-cell tumors, and retains some of the properties of the cell from which it develops. Some of these tumors represent massive outgrowth of a rare cell type, for example the lymphoid progenitor cell. Two T-cell-related tumors are also included in this figure: thymomas derive from thymic stromal or epithelial cells, whereas the malignant cell in Hodgkin's disease is thought to be an antigen-presenting cell. Some characteristic cell-surface markers for each stage are also shown. CD10 is also known as common acute lymphoblastic leukemia antigen or CALLA and is a widely used marker for C-ALL. Note that T-cell chronic lympho-cytic leukemia (CLL) cells express CD8, whereas the other T-cell tumors mentioned express CD4.

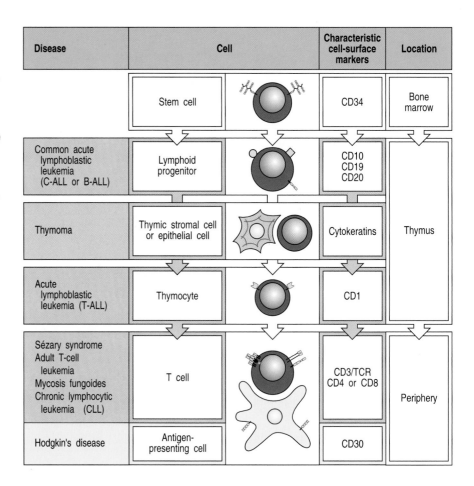

Disease	Cell		Characteristic cell-surface markers	Location
	Stem cell		CD34	Bone marrow
Common acute lymphoblastic leukemia (C-ALL or B-ALL)	Lymphoid progenitor		CD10 CD19 CD20	
Thymoma	Thymic stromal cell or epithelial cell		Cytokeratins	Thymus
Acute lymphoblastic leukemia (T-ALL)	Thymocyte		CD1	
Sézary syndrome Adult T-cell leukemia Mycosis fungoides Chronic lymphocytic leukemia (CLL)	T cell		CD3/TCR CD4 or CD8	Periphery
Hodgkin's disease	Antigen-presenting cell		CD30	

distinctive markers of the T-cell developmental lineage, such as CD25, but do not express any of the three cell-surface markers that define mature T cells, namely the T-cell receptor complex (detected as CD3), CD4, or CD8. Because they lack CD4 and CD8, such cells are called double-negative thymocytes.

Some double-negative cells (representing about 20% of all double-negative cells in the thymus) express the genes encoding the rare $\gamma{:}\delta$ receptor; these cells represent a separate developmental lineage of CD4⁻ CD8⁻ thymocytes. A second double-negative population expresses the pre-T receptor; these cells mature into double-positive (CD4⁺CD8⁺) thymocytes. Once the mature T-cell receptor α chain is expressed, these rapidly proliferating cells phosphorylate RAG-2 protein and degrade it; consequently there is no further gene arrangement.

Small double-positive thymocytes express only low levels of the T-cell receptor, and most (more than 95%) are destined to die. These cells express receptors that cannot recognize self MHC and thus fail positive selection in the thymus. Double-positive cells that do recognize self MHC, however, mature to express high levels of T-cell receptor and subsequently cease to express one or other of the two co-receptor molecules, becoming either CD4 or CD8 single-positive thymocytes. Thymocytes also undergo negative selection during the double-positive stage. Those that survive this dual screening mature into single-positive T cells that are rapidly exported from the thymus to join the peripheral T-cell repertoire (see Figs. 27.2 and 5.3).

This case concerns a child with a malignant monoclonal tumor of double-positive immature T cells with abnormal function.

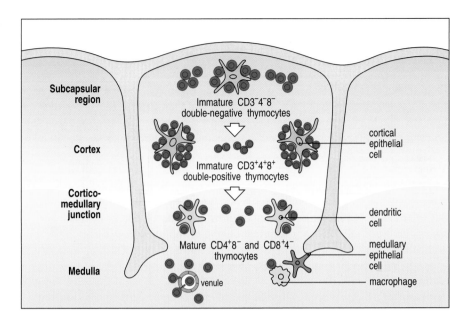

Fig. 27.2 Thymocytes go through several distinct developmental stages, which can be distinguished by the presence or absence of certain cell-surface proteins. The earliest cells to enter the thymus are double-negative thymocytes that lack the T-cell receptor (detected as CD3), and the co-receptors CD4 and CD8. As these cells proliferate and mature into double-positive thymocytes, bearing CD3 and both CD4 and CD8, they move deeper into the thymic cortex. Finally, the medulla contains only mature single-positive T cells, which eventually leave the thymus and enter the bloodstream.

The case of Elizabeth Bennet: the consequence of unrestrained growth of a cytokine-producing T-cell clone.

Elizabeth Bennet was a healthy 8-year-old when she developed chest pain, a cough, and shortness of breath. After 3 months of these symptoms, her parents sought advice from their family doctor.

On physical examination Elizabeth was found to be thin (weighing 28 kg) and 129 cm tall. She had a bulging left chest, no sound of breath over the left chest, and decreased breath sounds over the right chest. Cervical, axillary, and left inguinal lymph nodes were moderately enlarged. The liver and spleen were enlarged; their edges were each 5 cm below the costal margin.

Her chest radiograph revealed a completely opaque left hemithorax and a mediastinal shift to the right. A computed axial tomography scan (CAT scan) showed a solid tumor filling the left hemithorax. Pleural fluid obtained from the left chest contained abundant eosinophils and few lymphocytes but no malignant cells. A biopsy of the tumor showed that 98% of the tumor cells expressed CD3, CD4, and CD8; this suggested that the tumor was derived from the double-positive $CD4^+CD8^+$ population of thymocytes.

On admission to the Children's Hospital, Elizabeth's leukocyte count was 11,400 cells μl^{-1} (slightly elevated compared with normal), with 67% eosinophils, 31% neutrophils, 7% lymphocytes, and 5% monocytes. Bone marrow aspiration revealed a hypercellular marrow with increased eosinophil precursors. No malignant cells were observed in peripheral blood or bone marrow smears.

Because of the blood and bone marrow eosinophilia, serum IgE level was measured and was found to be >10,000 IU ml^{-1}. There was no personal or family history of allergic disease. A battery of 24 hypersensitivity skin tests for inhalant allergens

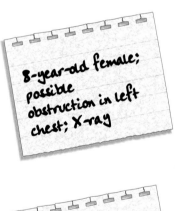

8-year-old female; possible obstruction in left chest; X-ray

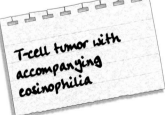

T-cell tumor with accompanying eosinophilia

Anti-tumor treatment; IgE levels decreased to normal

was negative, as were repeated stool examinations for parasites and their eggs. The IgE was polyclonal in origin. IgG was 1388 mg dl^{-1} (normal 568–1100), IgA was 123 mg dl^{-1} (normal 57–414), and IgM was 84 mg dl^{-1} (normal 120–274).

Eight weeks after treatment with cytotoxic drugs was begun there was no clinical or radiologic evidence of the tumor. Serum IgE at that time was 767 IU ml^{-1}. Serum IgM, IgG, and IgA remained essentially unchanged. Five months later there was still no evidence of disease and Elizabeth's serum IgE level at that time was 200 IU ml^{-1}.

Two months later a mild enlargement of the left side of the mediastinum was observed on a chest radiograph. Subsequently, over a 3-month period, a left mediastinal mass became clearly visible and grew larger, despite the continued administration of cytotoxic drugs. Tumor growth was accompanied by a rise in serum IgE (up to 3900 IU ml^{-1}). A further course of anti-tumor drugs—cytosine arabinoside and VP16—was administered. This was quickly followed by a decrease in serum IgE, from 3900 to 746 IU ml^{-1} over a period of 8 days.

The peripheral blood eosinophilia that had initially been present rapidly resolved after treatment with prednisone. However, with the recurrence of the tumor mass the blood eosinophil count rose rapidly. Although the neutrophil count rose 2.5-fold at that time, the eosinophil count rose concomitantly from 100 to 6800 μl^{-1}. Supernatants from cultures of the tumor cells contained interleukin (IL)-4 and induced IgE synthesis when added to cultures of normal peripheral blood mononuclear cells. IgE synthesis was blocked by anti-IL-4 neutralizing monoclonal antibody. The same supernatants induced the differentiation of cord blood mononuclear cells into eosinophils after 3 weeks in culture. Differentiation was blocked by anti-IL-5 neutralizing monoclonal antibody.

T-cell lymphoma.

The high division rate of maturing lymphocytes together with the activity of DNA recombination mechanisms responsible for generating antigen receptor diversity provide a fertile ground for malignant transformation and the generation of T-cell lymphomas. Even terminally differentiated T cells retain the capacity to proliferate in response to appropriate stimuli and therefore retain potentially lifelong capability for neoplasia. Once transformed, these T cells or their developmentally arrested precursors frequently retain the pattern of surface antigen expression and function of non-transformed lymphocytes at the same developmental stage.

Neoplastic transformation at various stages of T-cell development gives rise to leukemias and lymphomas with distinct clinical characteristics. Common acute lymphoblastic leukemias (C-ALL or B-ALL) are derived from early lymphoid precursors and do not express CD3, CD4, CD8, or any of the rearranged T-cell receptor genes. T-cell acute lymphoblastic leukemia (T-ALL) and lymphoblastic lymphoma can result from the transformation of any of the stages of T-cell differentiation including double-negative (CD4$^-$CD8$^-$) and double-positive (CD4$^+$CD8$^+$) precursors or mature single-positive (CD4$^+$or CD8$^+$) T cells. Neoplasms of mature CD4 helper T cells include large-cell lymphoma, adult T-cell leukemia/lymphoma (induced by the retrovirus

HTLV-1), cutaneous T-cell lymphoma or Sézary syndrome, and hairy-cell leukemia. Transformation of the mature CD8 single-positive phenotype is observed in some large-cell lymphomas.

A wide variety of immunologic abnormalities can be observed in patients with lymphoma and these reflect the functional effects of the transformed cells. Some T-cell lymphomas retain function: in particular, the cells of cutaneous T-cell lymphomas (Sézary syndrome) can display helper activity and can induce normal B cells to undergo isotype switching and produce immunoglobulin. Unlike normal T cells, however, the transformed Sézary syndrome lymphocytes perform these helper functions without expressing interleukin (IL)-2 receptors and do not require IL-2 for their ongoing proliferation.

Another setting in which the transformation of mature CD4$^+$ helper T cells occurs is in diseases associated with human T-cell leukemia/lymphoma virus (HTLV). These malignancies, also referred to as adult T-cell leukemia (ATL), are particularly prevalent in parts of Japan. They are induced by the HTLV-1 retrovirus; the cells express CD3 and CD4 antigens but not CD8, consistent with a mature helper T cell phenotype. Unlike other T-cell malignancies, these HTLV-1-transformed lymphocytes express high levels of IL-2 receptors and require IL-2 for their proliferation and function.

The lymphoma described in Elizabeth represents a case of transformation of cells with some immunological function. Its pattern of surface antigen expression—CD3$^+$CD4$^+$ and CD8$^+$—is reminiscent of the double-positive cells that constitute most of the thymus and distinguishes this lymphoma from the Sézary syndrome and HTLV-1 neoplasms discussed above. The double-positive thymocytes are in the process of undergoing selection on the basis of their T-cell receptor specificity before maturing to a single-positive CD4$^+$ or CD8$^+$ phenotype and being exported to the peripheral immune organs. Typically, such cells are not found in the periphery and are not known to secrete cytokines or carry out immune effector functions. This double-positive lymphoma was clearly shown to secrete both IL-4 and IL-5, properties usually associated with CD4$^+$ T cells differentiated to the T$_H$2 phenotype.

Discussion and questions.

1 *Elizabeth's tumor cells secreted IL-4 and IL-5. What other cytokines could be secreted by the tumor that would also cause eosinophilia and elevation of serum IgE levels?*

IL-13 has activities that overlap with IL-4. The two molecules, which are both produced by activated CD4$^+$ or CD8$^+$ T cells, are structurally very similar and probably arose by gene duplication. Both cytokines promote IgE synthesis, B-cell proliferation, and MHC class II gene expression. There are, however, some important differences between them. IL-13 seems to be produced at similar levels by helper T cells of the T$_H$0, T$_H$1, and T$_H$2 types and is produced very rapidly after T-cell activation, whereas IL-4 is primarily the product of T$_H$2 cells and appears later in the response. IL-13 does not have the T$_H$2-promoting activity of IL-4 and, unlike IL-4, does not induce T-cell proliferation. IL-13 promotes both eosinophil survival and recruitment.

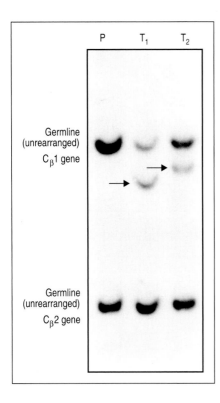

Fig. 27.3 The unique rearrangement events in each T cell can be used to identify tumors of T cells. Tumors are the outgrowth of a single transformed cell. Thus, each cell in a tumor has an identical pattern of rearranged T-cell receptor genes. This figure shows the gel electrophoresis of DNA fragments containing the T-cell receptor β-chain constant regions; the DNA is obtained either from the placenta (lane P), a tissue in which the T-cell receptor genes are not rearranged, or from peripheral blood lymphocytes from two patients suffering from T-cell tumors (lanes T_1 and T_2). Bands corresponding to the unrearranged $C_\beta 1$ and $C_\beta 2$ genes can be seen in all lanes. Additional bands corresponding to specific rearrangements (arrowed) can be seen in each of the tumor samples, indicating that a large proportion of the cells in the sample carry an identical gene rearrangement. Note that these are the only discrete additional bands that can be seen in these samples; no bands deriving from rearranged genes in the normal lymphocytes also present in the patient's serum can be seen, as no one rearranged band is present at sufficient concentration to be detected in this assay. Photograph courtesy of T Diss.

---

2 *Elizabeth's serum IgG was elevated. How do you explain this?*

IL-4 induces isotype switching to IgE and IgG4 in human B cells. In addition, IL-4 promotes T-cell differentiation to the T_H2 phenotype. T_H2 cells produce more IL-4, further enhancing humoral immune responses. In addition, T_H2 cells make IL-5, IL-6, IL-10, and IL-13, which promote B-cell expansion and immunoglobulin synthesis.

3 *What definitive test will establish that Elizabeth's tumor represents a monoclonal expansion of a T-cell clone?*

The demonstration that the tumor cells all express the same T-cell receptor. Proliferation of T cells can arise either as the result of clonal malignant transformation (leukemia/lymphoma) or from polyclonal expansion driven either by antigen or as the result of dysregulated proliferation. Antigen-driven T cells are polyclonally expanded. As a result, they express a heterogeneous population of T-cell receptors of various specificities at their cell surface. These receptors are constructed during T-cell development by random splicing of genomic V, D, and J segments. As neoplasms are clonal, T-cell leukemias and lymphomas express only a single T-cell receptor (Fig. 27.3).

4 *Administration of corticosteroids (eg prednisone) results in a rapid decrease in levels of circulating and tissue eosinophils (which is also commonly seen in allergic conditions after corticosteroid treatment). Propose a mechanism for this effect.*

Corticosteroids are known to inhibit the transcription of a number of cytokines associated with allergic states, including IL-4, IL-5, and GM-CSF. Corticosteroids function by interacting with cytosolic receptors, inducing a conformational change in these receptors that enables the steroid–receptor complex to act as a transcription factor and bind to specific sequences known as glucocorticoid response elements (GREs) in the promoters of certain genes in the nucleus (Fig. 27.4). Among the genes containing GREs in their promoters is the anti-inflammatory gene family IκBα. Normally, the IκB protein binds to NFκB transcription factors in the cytosol, preventing their translocation to the nucleus where they promote the transcription of a wide variety of pro-inflammatory genes (including that for GM-CSF). After the corticosteroid-induced upregulation of IκB, NFκB-regulated gene transcription is suppressed and the production of a number of pro-inflammatory cytokines is reduced.

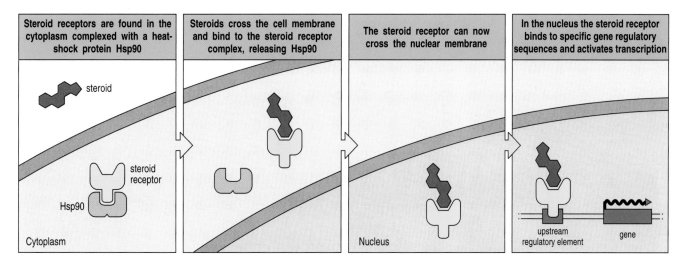

| Steroid receptors are found in the cytoplasm complexed with a heat-shock protein Hsp90 | Steroids cross the cell membrane and bind to the steroid receptor complex, releasing Hsp90 | The steroid receptor can now cross the nuclear membrane | In the nucleus the steroid receptor binds to specific gene regulatory sequences and activates transcription |

Fig. 27.4 Mechanisms of steroid action. Corticosteroids are lipid-soluble molecules that enter cells by diffusing across the plasma membrane and bind to their receptors in the cytosol. The binding of corticosteroid to the receptor displaces a dimer of a heat-shock protein named Hsp90, exposing the DNA-binding region of the receptor. The receptor–steroid complex then enters the nucleus and binds to specific DNA sequences in the promoter regions of steroid-responsive genes. Corticosteroids exert their effects by modulating the transcription of a wide variety of genes.

5 | *Besides CD4$^+$ T$_H$2 cells, what other thymus-derived population of cells represents a potential source of IL-4?*

Although CD4 T$_H$2 lymphocytes represent a major source of IL-4, other cells have been reported to produce this cytokine. In mice, a unique population of thymocytes bearing the cell-surface antigen NK1.1 may represent a major source of IL-4 early on in immune responses. NK1.1$^+$ T cells express an unusually restricted pattern of α:β T-cell receptors (V$_\alpha$24, J$_\alpha$Q7:V$_\beta$11) and respond to antigen-presenting cells bearing the MHC class I-like molecule CD1 (Fig. 27.5). Unlike CD4 T$_H$2 cells, they release large amounts of IL-4 within 30–90 minutes of primary stimulation. In addition to these thymus-derived cells, mast cells and basophils have also been reported to produce IL-4.

Transgenic mice overexpressing the NK1.1-type T-cell receptor have a marked tendency towards the production of IL-4 and IgE, suggesting a potential role for this population of cells in allergy. However, mice in which the β$_2$-microglobulin gene has been knocked out, and that therefore cannot express CD1 and are deficient in NK1.1$^+$ thymocytes, develop normal IgE and IL-4 responses to allergic stimuli and parasitic infection. This indicates that NK1.1$^+$ T cells are not essential for such responses.

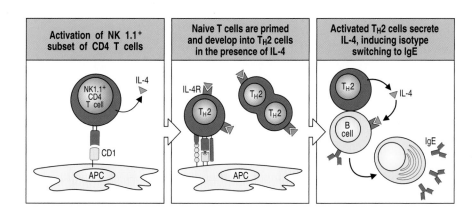

| Activation of NK 1.1$^+$ subset of CD4 T cells | Naive T cells are primed and develop into T$_H$2 cells in the presence of IL-4 | Activated T$_H$2 cells secrete IL-4, inducing isotype switching to IgE |

Fig. 27.5 NK1.1$^+$ T cells produce IL-4 on stimulation. IL-4 is secreted early in some immune responses by a small subset of CD4 T cells (NK1.1$^+$ CD4 T cells) that interact with antigen-presenting cells bearing the non-classical MHC class I-like molecule CD1. Naive T cells being primed by their first encounter with antigen are driven to differentiate into T$_H$2 cells in the presence of this early burst of IL-4. These mechanisms have been characterized in mice. It is not yet known whether the same pathways operate in humans.

CASE 28 | Interferon-γ Receptor Deficiency

The destruction of intracellular microorganisms in macrophages.

Certain pathogens such as mycobacteria, *Listeria*, *Leishmania*, and *Salmonella* take up residence in macrophages and are thereby protected from elimination by antibody or cytotoxic T cells. These microorganisms can be eliminated only when their host macrophages are activated and produce increased amounts of nitric oxide, oxygen radicals, and other microbicidal molecules (Fig. 28.1). The activation of macrophages is masterminded by T_H1 cells (Fig. 28.2); the most important cytokine involved in macrophage activation is interferon (IFN)-γ.

Topics bearing on this case:
Macrophage activation
Microbicidal action of phagocytes
Cytokine receptor signaling pathways

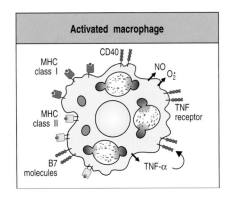

Fig. 28.1 Activated macrophages undergo changes that greatly increase their antimicrobial effectiveness and amplify the immune response. Once the macrophage is activated, lysosomes fuse with the intracellular vesicles within which the pathogenic bacteria (red) reside, which exposes the microorganisms to degradative enzymes and other microbicidal agents. Activated macrophages also increase their expression of receptors for tumor necrosis factor (TNF), and secrete TNF-α. This autocrine stimulus synergizes with interferon (IFN)-γ secreted by T$_H$1 cells to increase the antimicrobial action of the macrophage, in particular by inducing the production of nitric oxide (NO) and oxygen radicals (O·$_2$). The macrophage also increases the expression of CD40, which by interaction with the CD40 ligand on T cells upregulates the expression of B7 proteins and increases the expression of class II MHC molecules on the macrophage, thus allowing the further activation of resting CD4 T cells.

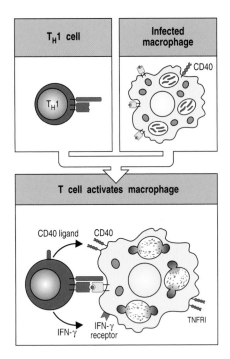

Fig. 28.2 T$_H$1 cells activate macrophages to become highly microbicidal. When a T$_H$1 cell specific for a bacterial peptide contacts an infected macrophage, the T cell is induced to secrete the macrophage-activating factor IFN-γ by IL-12 and to express CD40 ligand. Together, these newly synthesized T$_H$1 proteins activate the macrophage.

When macrophages take up microorganisms by phagocytosis, they secrete interleukin (IL)-12. This cytokine is necessary for the induction of IFN-γ synthesis by T cells and natural killer (NK) cells. IL-12 furthermore favors the maturation of T$_H$1 cells, which activate macrophages, and suppresses the maturation of T$_H$2 cells, which secrete a cytokine, IL-10, involved in the deactivation of macrophages.

IFN-γ acts at a receptor on the macrophage surface. This receptor is composed of two different types of polypeptide chain—IFN-γ receptor 1 and IFN-γ receptor 2, each of which is associated with a tyrosine kinase—Jak1 and Jak2 respectively. When a dimer of IFN-γ binds to two molecules of the IFN-γ receptor 1, it causes them to associate with two IFN-γ receptor 2 chains (Fig. 28.3) and this initiates a signaling pathway that eventually results in changes in gene transcription.

As we shall see in this case, a mutation in the gene encoding the IFN-γ receptor 1 chain has dramatic effects on the capacity to fight certain pathogens.

The case of Clarissa Dalloway: a relentless infection due to mycobacteria.

Clarissa Dalloway was the first child born to a couple who lived in an isolated fishing village on the coast of Maine. Her parents thought that they were distantly related to each other. The fishermen of this village were all descended from English settlers who came there in the late 17th century and there was much intermarriage in the community.

Clarissa was well at birth and developed normally until she was around 2½ years old. Her mother then noticed that she was not eating well, had diarrhea, and was losing weight. She took Clarissa to a pediatrician in the nearest town of Bath, Maine. The pediatrician, Dr Woolf, noted enlarged lymph nodes and ordered an ultrasound and CT scan of the chest and abdomen. These showed enlarged lymph nodes in the mesentery and para-aortic region and Dr Woolf referred Clarissa to the Children's Hospital in Boston.

Blood tests revealed a white blood cell count of 9400 μl^{-1}, of which 55% were neutrophils, 30% lymphocytes, and 15% monocytes (slightly elevated). Her serum IgG was 1750 mg dl^{-1}, IgA 450 mg dl^{-1}, and IgM 175 mg dl^{-1} (these immunoglobulin values are all elevated).

2½-year-old female child with enlarged lymph nodes. Order blood test.

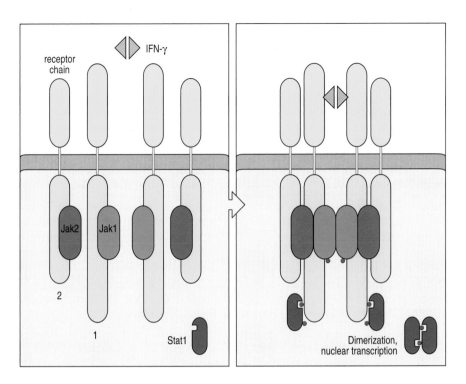

Fig. 28.3 Diagram of the IFN-γ receptor and its activation of the JAK–STAT signaling pathway. A dimer of IFN-γ causes clustering of the four chains of the IFN-γ receptor. The receptor 1 chains become phosphorylated by Jak1, forming a site to which the transcription factor Stat1α can bind. This in turn becomes phosphorylated, dimerizes, and is then transported to the nucleus, where it upregulates the various genes that produce the activated macrophage phenotype. The IFN-γ receptor 1 chain is encoded on chromosome 6, whereas the receptor 2 chain is encoded on chromosome 21.

It was decided to biopsy the enlarged lymph nodes. On histological examination they exhibited marked proliferation of histiocytes, and many neutrophils were seen in the lymph node. There was no granuloma formation and no giant cells were seen. An acid-fast stain for mycobacteria (see, for example, Fig. 29.4) revealed numerous microorganisms, and *Mycobacterium avium intracellulare* was cultured from the lymph nodes as well as from the blood.

Despite appropriate antibiotic treatment for the mycobacterial infection, Clarissa eventually developed infiltrates in the lungs and progressive enlargement of the spleen. At 6 years old she developed sepsis and *Salmonella paratyphi* was cultured from her blood. She was successfully treated with antibiotics for this infection but soon after she developed meningitis and died. *M. avium intracellulare* was cultured from the cerebrospinal fluid.

A detailed family history revealed that Clarissa had three male cousins, two of whom were brothers, who had died of mycobacterial infections. In one case *M. chelonei* had been cultured from lymph nodes; *M. fortuitum* and *M. avium intracellulare* had been cultured from blood and lymph nodes of the two brothers.

do lymph node biopsy

Interferon-γ receptor deficiency.

It is estimated by the World Health Organization that 1.7 billion human beings currently alive have been infected with mycobacteria. Yet only a very small fraction of these individuals develop disease due to this infection. The AIDS epidemic has dramatically increased the incidence of mycobacterial disease due to *M. tuberculosis* and to other mycobacterial species such as *M. avium intracellulare*, which are collectively called 'atypical mycobacteria.' In general, atypical mycobacteria do not cause disease in immunologically normal human beings, except for swollen lymph nodes in which these mycobacteria survive. Infection with atypical mycobacteria causes a positive tuberculin reaction to develop.

Although there are many ethnic differences in susceptibility to *M. tuberculosis*, no single genetic factor had ever been found in humans to explain this susceptibility until Clarissa Dalloway and her family were found to have a genetic defect in the IFN-γ receptor 1 gene. The genetic defect was ascertained in this family by mapping the susceptibility gene to chromosome 6q22, the map location of the receptor gene.

This finding prompted the examination of children who had developed progressive mycobacterial infection after immunization with the BCG (Bacille Calmette-Guérin) vaccine, and who did not have severe combined immunodeficiency (see Case 5). More kindreds with the IFN-γ receptor 1 deficiency were discovered as a result of this lead. Subsequently, a similar susceptibility to mycobacterial infection has been found in patients with defects in IL-12 synthesis or in the IL-12 receptor. Thus the dependence of IFN-γ synthesis on IL-12 has been neatly confirmed by these human mutations.

Discussion and questions.

1 *How do you explain the alarming increase in the incidence of mycobacterial infections in patients with AIDS?*

Mycobacteria, particularly atypical mycobacteria, are ubiquitous in the environment and any infection is normally contained by T-cell action. AIDS, however, is characterized by a marked reduction in the number of CD4 T_H1 cells. Hence the production of IFN-γ is compromised and patients with AIDS have difficulty in activating their macrophages and clearing mycobacteria.

2 *Clarissa Dalloway and her cousins had positive tuberculin skin tests. Would you have predicted this?*

The delayed-type hypersensitivity reaction is provoked by a few T cells that are specific for tuberculin. After binding antigen, these T cells secrete chemokines that non-specifically recruit macrophages and other inflammatory cells. The secretion and action of these chemokines is not dependent on IFN-γ. It is therefore not surprising that Clarissa and her cousins developed positive tuberculin skin tests.

3 *Clarissa and her cousins also did not develop granulomas as a result of infection with atypical mycobacteria. On one occasion mycobacteria were surprisingly cultured from Clarissa's blood. What do these observations tell us?*

Granulomas form where there is local persistence of antigen, antigen-specific T cells, and activated macrophages. As these children could not activate their macrophages it is not surprising that they did not form granulomas. Granulomas can be beneficial in that they wall off and prevent the spread of microorganisms. These children could not do that and we find the myco-bacteria spreading via the blood stream—a highly unusual finding in mycobacterial disease.

4 *Clarissa also had a problem with salmonella, but she had no problem with pneumococcal infection or with any viruses, such as chickenpox. How do you explain this?*

Infection with pyogenic bacteria such as pneumococci is controlled and terminated by antibody and complement. In fact Clarissa had a very high level of IgG (1750 mg dl^{-1}), probably as a result of increased IL-6 production induced by the chronic mycobacterial infection. Most viral infections, such as chickenpox, are terminated by cytotoxic CD8 cells. Activation of these cells is not dependent on IFN-γ. However, there is evidence from mice that a lack of the IFN-γ receptor increases susceptibility to certain viruses, including vaccinia virus and lymphocytic choriomeningitis virus. Salmonella take up residence as an intracellular infection of macrophages and become inaccessible to antibody and complement; they can be destroyed in this site only when macrophages are activated by IFN-γ.

CASE 29 | Lepromatous Leprosy

T$_H$1 versus T$_H$2 responses in the outcome of infection.

Mature CD4 T cells emerging from the thymus differentiate into CD4 T cells of two different phenotypes—T$_H$1 and T$_H$2—with different functions in the immune response. These T$_H$1 and T$_H$2 cells develop from less mature effector cells, designated T$_H$0 (Fig. 29.1). The two types of T cell are distinguished chiefly by the cytokines that they secrete when activated. T$_H$1 cells secrete interleukin-2 (IL-2), interferon-γ (IFN-γ), and tumor necrosis factor-β (TNF-β); T$_H$2 cells, often called helper T cells, secrete IL-4, IL-5, and IL-10 (Fig. 29.2).

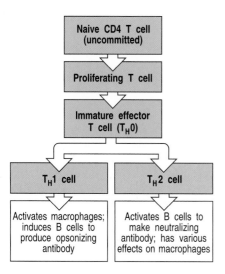

which has some of the effector functions characteristic of T_H1 and T_H2 cells. The T_H0 cell has the potential to become either a T_H1 or a T_H2 cell. T_H0 cells might also have some effector actions in their own right.

The consequences of the decision to differentiate into T_H1 or T_H2 cells are profound, as selective production of T_H1 cells biases the immune response towards macrophage activation and cell-mediated immunity, whereas selective production of T_H2 biases the response towards antibody production and humoral immunity. The decision as to which pathway a T_H0 cell will follow is made during its first encounter with antigen. An antigen that interacts strongly with the T-cell receptor causes the T_H0 cell to mature into a T_H1 cell, whereas a weak interaction leads to T_H2 development. Differentiation is also dependent on cytokines. T_H1 differentiation is dependent on IL-12 and IFN-γ, whereas T_H2 differentiation is dependent on IL-4 (see Fig. 29.2). These cytokines trigger pathways of signal transduction; for example, mice deficient in the intracellular signaling molecule Stat6, which is induced by IL-4, fail to develop T_H2 cells.

Other factors influencing the T-cell phenotype are the amount of antigen present and thus which cells are most likely to present it. Large amounts of antigen are usually presented by dendritic cells, which produce IL-12 and therefore favor T_H1 differentiation. A limited amount of antigen leads to preferential presentation by antigen-specific B cells that take up antigen more avidly; they induce T_H2 differentiation. The co-stimulatory molecules (B7.1 versus B7.2) expressed by the antigen-presenting cells also influence the maturation process in that B7.1 (CD80) is more likely to provoke T_H1 development, and B7.2 (CD86) to provoke T_H2 development.

Because the decision to differentiate into T_H1 versus T_H2 cells occurs early in an adaptive immune response, the ability of pathogens to stimulate cytokine production by cells of the innate, non-adaptive immune system has an important role in determining the subsequent course of the response. Infectious agents that invade or non-specifically activate macrophages and NK cells, as do most viruses and intracellular bacteria such as mycobacteria, induce cells to secrete IL-12, thus favoring the differentiation of T_H1 cells, which secrete IFN-γ. This loop is amplified because IFN-γ in turn favors T_H1 development and blocks the development of T_H2 cells (see Fig. 29.2). IL-12 also enhances the proliferation of T_H1 cells but has no effect on T_H2 cells because they lack the β chain of the IL-12 receptor and therefore do not respond to the mitogenic effects of IL-12.

The differentiation of T_H2 cells is favored by pathogens, such as parasites, that elicit IL-4 production from specialized subsets of cells that include mast cells, eosinophils, and thymic-derived T cells expressing both the NK1.1 marker and a T-cell receptor of restricted $V_β$ and invariant $V_α$ chain usage. This loop is amplified by the cytokines produced by T_H2 cells—IL-4 and IL-10. IL-4 promotes the development of T_H2 cells and IL-10 blocks T_H1 development (see Fig. 29.2).

Once one T_H phenotype becomes dominant in the course of a response, it is difficult to shift the antigen-specific response to the other. One reason for this is that the cytokine products of T_H1 and T_H2 cells are reciprocally inhibitory (see Fig. 29.2). The outcome of certain infections is greatly influenced by the type of T-cell response elicited. As we see in this case, infection with *Mycobacterium leprae*, the leprosy bacillus, is a good example.

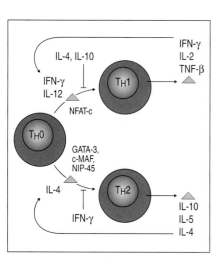

Fig. 29.2 Differentiation of T_H0 cells into either T_H1 or T_H2 cells depends on the cytokines present. IL-4 induces T_H2 differentiation, whereas IL-12 and IFN-γ induce differentiation into T_H1 cells. IL-4, which is secreted by T_H2 cells, also inhibits T_H1 development. Similarly, cytokines produced by T_H1 cells, for example IFN-γ, inhibit T_H2 differentiation. T_H1 or T_H2 development is driven by transcription factors induced by the cytokines. Induction of the transcription factors GATA-3, c-MAF and NIP-45 in naive T cells leads to T_H2 development, whereas induction of NFAT-c leads to T_H1 development.

The case of Ursula Iguaran: a T$_H$2 response to the leprosy bacillus has severe consequences.

Ursula first sought medical advice when she was 18 years old, having left her home in Colombia to attend Harvard University on a scholarship. From the age of 16, she had started to notice a gradual loss of sensation on the backs of her hands and had developed hypopigmented lesions over both arms. The lesions progressively became worse and she noticed that she was losing her eyelashes and hairs from her eyebrows. She also experienced recurrent nose bleeds. A month after first noticing the hair loss she decided to seek medical help.

18-year-old female; light-colored lesions on skin

On examination at the physician's office, Ursula seemed to be well apart from her immediate symptoms. She reported a history of mild asthma, which required treatment with inhaled β$_2$-adrenergic agents on an as-needed basis. Multiple hypopigmented macules (coin-sized raised lesions with ill-defined borders) were evident on her skin, along with cutaneous nodules 1 cm in diameter. These lesions were predominantly on her elbows, wrists, and hands (Fig. 29.3) and showed traces of dried blood; she also had similar lesions on her knees, ears, and buttocks. The absence of eyelashes and the ends of her eyebrows was obvious.

Acid-fast bacilli (? mycobacteria) in skin lesions; leprosy?

Cardiovascular and abdominal examinations were normal; a neurological examination was negative except for a decreased response to pinprick on the outer edges of the right and left hand and the right fourth and fifth fingers. There was a flexion contracture of the fourth and fifth fingers of both hands so that she could not straighten these fingers completely.

On blood test, her hematocrit was 35.1%; white blood count was 7100 μl^{-1}, with 68% neutrophils, 23% lymphocytes, 5% monocytes, and 4% eosinophils (all normal values). Serum electrolytes were normal. Because her symptoms were becoming more severe, she was referred to a dermatologist. She told him that she had grown up in a small village on the Caribbean coast of Colombia where many people, including her mother, had leprosy. The dermatologist performed a biopsy of the lesions on her left arm and right forearm, which revealed numerous acid-fast bacilli in clumps. A routine hematoxylin and eosin stain of lesion tissue showed up numerous Virchow's cells (highly vacuolated cells of the macrophage lineage also known as foam cells) and few lymphocytes (Fig. 29.4). Cultures for acid-fast bacilli were negative.

The suspected diagnosis of lepromatous leprosy led to a more extensive immunologic work-up. Delayed hypersensitivity skin tests with intradermal injections of candida, mumps, and tuberculin antigens showed no reactions when the injection sites were inspected 48 and 72 hours later. Ursula's serum IgG was mildly elevated at 1800 mg dl^{-1} (normal 600–1100 mg dl^{-1}); her IgA and IgM levels were normal.

A diagnosis of lepromatous leprosy was made on the basis of the presence of acid-fast bacilli in the biopsy and Ursula's progressive neurologic symptoms. She was placed on a multiple drug regime consisting of dapsone, clofazamine, and rifampin drugs that kill *M. leprae*. Her skin lesions gradually flattened and improved.

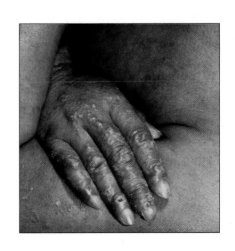

Fig. 29.3 Cutaneous nodules in lepromatous leprosy. Patients with lepromatous leprosy have multiple skin lesions. This photograph shows subcutaneous nodules on the hand.

Fig. 29.4 Responses to _M. leprae_ are sharply differentiated in lepromatous and tuberculoid leprosy. The photographs show sections of lesion biopsies stained with hematoxylin and eosin. Infection with _M. leprae_ bacilli, which can be seen in the right-hand photograph as numerous small dark red dots inside macrophages, can lead to two very different forms of the disease. In tuberculoid leprosy (left), growth of the microorganism is well controlled by T_H1-like cells that activate infected macrophages. The tuberculoid lesion contains granulomas (see Case 15) and is inflamed, but the inflammation is localized and causes only local peripheral nerve damage. In lepromatous leprosy (right), infection is widely disseminated and the bacilli grow uncontrolled in macrophages. In the late stages there is severe damage to connective tissues and to the peripheral nervous system. There are several intermediate stages between these two polar forms. Photographs courtesy of G Kaplan.

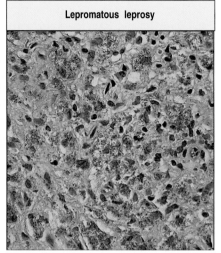

Infection with _Mycobacterium leprae_ can result in different clinical forms of leprosy	
There are two polar forms, tuberculoid and lepromatous leprosy, but several intermediate forms also exist	
Tuberculoid leprosy	**Lepromatous leprosy**
Organisms present at low to undetectable levels	Organisms show florid growth in macrophages
Low infectivity	High infectivity
Granulomas and local inflammation. Peripheral nerve damage	Disseminated infection. Bone, cartilage, and diffuse nerve damage
Normal serum immunoglobulin levels	Hypergammaglobulinemia
Normal T-cell responsiveness. Specific response to _M. leprae_ antigens	Low or absent T-cell responsiveness. No response to _M. leprae_ antigens

Lepromatous leprosy.

The classical clinical feature in patients with leprosy is the association of cutaneous lesions, neuropathologic changes, and deformities. Leprosy is caused by _Mycobacterium leprae_, which colonizes macrophages and other host cells and multiplies within them. Mycobacteria within macrophages are protected from attack by antibody and can be eliminated only when their host macrophages are activated and produce increased amounts of nitric oxide, oxygen radicals, and other microbicidal molecules. _M. leprae_ grows best at 30°C and therefore lesions tend to appear on the extremities—the colder areas of the body—for example the hands, ears, and buttocks as in Ursula's case. Unlike _M. tuberculosis_, _M. leprae_ does not grow in culture.

The clinical symptoms of leprosy vary depending on the type of immune response to the mycobacteria. The clinical spectrum is typically divided into two polar forms, tuberculoid and lepromatous leprosy, although intermediate forms exist. Tuberculoid leprosy is associated with a vigorous cell-mediated (T_H1) response against the bacillus. This results in macrophage activation with efficient killing of intracellular mycobacteria, localized tissue damage, and usually a milder clinical picture. In the lepromatous form, the T_H1-cell mediated response is defective and a T_H2 response predominates; this leads to a vigorous but ineffective antibody response against *M. leprae* and dissemination of the bacilli to other sites in the body, which results in further tissue destruction and aggravation of the symptoms. The importance of T_H1-derived IFN-γ in containing mycobacterial infections is further illustrated by the observation that infants with genetic defects in the IFN-γ receptor die from disseminated mycobacterial infections (see Case 28).

Infection with *M. leprae* illustrates a situation in which the same microorganism, in different individuals, can trigger either a T_H1 or a T_H2 response. A T_H1 response predominates in tuberculoid leprosy, in which the mycobacteria are contained within well-circumscribed granulomas and propagate poorly, usually accompanied by subsequent minimal tissue damage. In contrast, a T_H2 response predominates in lepromatous leprosy, in which the mycobacteria propagate rapidly, with resulting extensive tissue damage. Analysis of mRNA isolated from lesions of patients with lepromatous and tuberculoid leprosy illustrates the cytokine patterns in the two forms of the disease. T_H2 cytokines (IL-4, IL-5, and IL-10) dominate in the lepromatous form, whereas T_H1 cytokines (IL-2, IFN-γ, and TNF-β) dominate in the tuberculoid form (Fig. 29.5).

The neurologic damage in leprosy has two main causes. It can arise from bacterial multiplication within Schwann cells—the cells that form the insulating myelin sheath around some nerve cell axons. Disruption of the myelin sheath interferes with the normal conduction of nerve impulses along the axon. In the tuberculoid form, nerve damage also arises from the formation of granulomas and inflammation of the tissue surrounding the nerve. The nerve damage results in dysfunctional nerve terminals, resulting in decreased sensation and eventually a loss of motor function.

The nose bleeds that Ursula experienced are common in leprosy. They are due to large numbers of *M. leprae* in the nasal tissue with extensive involvement of the nasal mucosa, leading to congestion and breakage of blood vessels.

The T_H2 response can influence the course of the infection in various other ways. By binding to mycobacterial antigens displayed on the surface of infected cells, antibodies to the leprosy bacillus can interfere with the action of cytotoxic CD8 T cells. CD8 T cells can, in addition to their cytolytic function, also respond to antigen by secreting cytokines. Patients with lepromatous

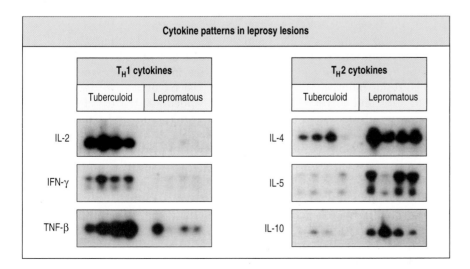

Fig. 29.5 Cytokine patterns in leprosy lesions. The cytokine patterns in the two polar forms of leprosy are distinctly different, as shown by Northern blot analysis of the mRNA from lesions of three patients with lepromatous leprosy and three patients with tuberculoid leprosy. Cytokine mRNAs typically produced by T_H2 cells predominate in the lepromatous form whereas cytokines produced by T_H1 cells predominate in the tuberculoid form. Cytokine blots courtesy of R L Modlin.

leprosy have CD8 T cells that suppress the T_H1 response by making IL-10 and TGF-β. IL-10 inhibits the development of T_H1 cells and inhibits both cytokine release from macrophages and their capacity to kill internalized microorganisms. TGF-β also inhibits macrophage intracellular killing capacity. Inhibition of macrophages leads to decreased production of IL-12, fewer T_H1 cells, and more T_H2 cells. In contrast, patients with the less destructive tuberculoid leprosy lack suppressor CD8 T cells and thus make a vigorous T_H1 response, leading to macrophage activation and the destruction of the leprosy bacilli.

Discussion and questions.

---

1 *Ursula did not respond to candida and mumps antigens, which are common recall antigens, with a delayed-type hypersensitivity reaction. Give a possible explanation.*

Absence of delayed-type hypersensitivity to a wide range of antigens unrelated to *M. leprae* is called anergy. This should not be confused with T-cell or B-cell anergy, although it might operate by similar mechanisms. In tuberculoid leprosy, there is a strong delayed-type hypersensitivity to *M. leprae* and no anergy. The existence of anergy in the lepromatous form of leprosy but not in the tuberculoid form is most probably due to the presence of regulatory CD8 T cells in lepromatous leprosy. The CD8 T cells secrete the cytokines IL-10 and TGF-β and thereby suppress antigen presentation by macrophages. These cytokines not only influence the T_H1 versus T_H2 phenotype, as discussed above, but can suppress T-cell responses to other unrelated antigens. IL-10 and TGF-β suppress not only the *M. leprae*-specific T cells, but also neighboring T cells, leading to global hyporesponsiveness, which was manifest in Ursula's case as anergy to candida and mumps antigens. However, Ursula's case is somewhat atypical; in many patients with lepromatous leprosy the unresponsiveness is confined to *M. leprae* and responses are made to other antigens. Other pathogens use the IL-10 pathway to produce anergy; the Epstein–Barr virus, for example, produces a protein, VIL-10, that is homologous to IL-10. Measles virus induces anergy by binding to CD46 on monocytes and inhibiting their production of IL-12.

2 *Which cytokine might be beneficial to a patient with lepromatous leprosy?*

The immune response in patients with lepromatous leprosy is skewed towards the T_H2 phenotype, leading to a disseminated infection. Because tuberculoid leprosy involves a T_H1 response and significantly reduced symptoms, we would like to switch the response to the T_H1 phenotype. Cytokines with the potential to inhibit T_H2 and induce a T_H1 response are IL-2, IFN-γ, and IL-12. Local injection of IFN-γ has been shown to lead to partial reversal of anergy and reduction of lesions. IFN-γ has also been shown to be effective in the treatment of other similar diseases, such as leishmaniasis. In the visceral form of leishmaniasis, the T-cell response is also skewed to the T_H2 phenotype. This is in contrast to the cutaneous form of leishmaniasis, which is accompanied by a T_H1 response. IL-12 might also be beneficial, as it can induce T_H1 cells and does not activate T_H2 cells.

$\boxed{3}$ *Describe the mechanism for Ursula's hypergammaglobulinemia.*

In lepromatous leprosy, a humoral immune response driven by T_H2 cells predominates, with vigorous antibody production, leading to hypergamma-globulinemia as observed in Ursula. The cytokines produced in T_H2 responses lead to enhanced immunoglobulin production. IL-4 induces isotype switching to IgE and increased production of IgG4 and IgE. IL-10 stimulates the production of IgG1 and IgG3, whereas IL-5 stimulates immunoglobulin production globally. Therefore, it is not uncommon to find hypergamma-globulinemia in patients who are producing a vigorous T_H2 response to an antigen.

$\boxed{4}$ *Why would Ursula be prone to asthma?*

Ursula's T_H2 response to *M. leprae* leads to increased production of IL-4 and IL-10. When she encounters a new antigen, her immune system will be awash with IL-4, triggering a T_H2 response to that antigen. This T_H2 response with its associated IL-4 production leads to an IgE response. Because asthma and other atopic diseases are T_H2-driven diseases involving IgE production, Ursula has a higher risk of developing asthma.

CASE 30 | Atopic Dermatitis

Skin as a target organ for allergy.

As we saw in Case 16, allergic or hypersensitivity reactions to otherwise innocuous antigens occur in certain individuals. The site of such reactions and the symptoms they produce will vary depending on the type of allergen and the route by which it enters the body. When inhaled or ingested allergens enter the bloodstream they can be carried to the skin, where they induce a chronic inflammation—known as atopic dermatitis—in sensitized (atopic) individuals who possess T cells and IgE specific for an allergen. Allergens that act by direct skin contact can also incite inflammation. The immunologic response of the skin to antigen is complex, resulting in inflammation and localized tissue destruction. This chronic inflammation is called eczema and is the most prominent symptom of atopic dermatitis.

The immunologic mechanisms underlying atopic dermatitis have much in common with other IgE-mediated allergies (type I hypersensitive reactions) such as allergic asthma. The T cells involved in atopic dermatitis are principally T_H2 cells, which in this case home to the skin. These and other inflammatory cells are recruited to the skin by the release of cytokines and chemokines from the affected tissue. Cytokines induced by the allergen, or as a result of scratching and damaging the skin, stimulate the keratinocytes, the epidermal skin cells, to produce chemokines and cytokines. Interleukin (IL)-1 and tumor necrosis factor (TNF)-α secreted by keratinocytes in turn stimulate dermal fibroblasts to secrete chemokines that are involved in recruiting T cells and inflammatory cells to the site of injury.

Topics bearing on this case:
IgE and allergic reactions
Inflammatory responses
Cytokines and chemokines
Preferential activation of T_H2 cells
Migration and homing of lymphocytes
Anti-inflammatory effects of corticosteroids
Bacterial superantigens

Fig. 30.1 Properties of selected chemokines. Chemokines fall into three related but distinct structural subclasses: CC chemokines have two adjacent cysteines at a particular point in the amino-acid sequence; CXC chemokines have the same two cysteine residues separated by another amino acid; and C chemokines (not shown) have only one cysteine residue at this site.

Chemokine class	Chemokine	Produced by	Receptors	Major effects
CXC	IL-8	Monocytes Macrophages Fibroblasts Keratinocytes Endothelial cells	CXCR1 CXCR2	Mobilizes, activates and degranulates neutrophils Angiogenesis
CC	MIP-1α	Monocytes T cells Mast cells Fibroblasts	CCR1, 3, 5	Competes with HIV-1 Anti-viral defense Promotes T_H1 immunity
	MIP-1β	Monocytes Macrophages Neutrophils Endothelium	CCR1, 3, 5	Competes with HIV-1
	MCP-1	Monocytes Macrophages Fibroblasts Keratinocytes	CCR2B	Activates macrophages Basophil histamine release Promotes T_H2 immunity
	RANTES	T cells Endothelium Platelets	CCR1, 3, 5	Degranulates basophils Activates T cells Chronic inflammation
	Eotaxin	Endothelium Monocytes Epithelium T cells	CCR3	Role in allergy

Chemokines are small polypeptides that are synthesized by many cells, including keratinocytes, fibroblasts, T cells, eosinophils, and macrophages (Fig. 30.1). They act through receptors that are members of the G-protein-coupled seven-span receptor family. In atopic dermatitis, activated keratinocytes and dermal fibroblasts secrete the chemokines RANTES, eotaxin, and MCP-4, which attract lymphocytes, eosinophils, and macrophages into the skin. The chemokines and receptors known to be involved in atopic dermatitis are listed in Fig. 30.2.

A chronic inflammatory reaction is sustained by the lymphocytes, eosinophils, and other inflammatory cells that are attracted out of the blood vessels at the site of inflammation. The first step in the process of lymphocyte homing to skin is the reversible binding (rolling) of lymphocytes to the

Fig. 30.2 Chemokines that act on T cells and eosinophils. IP-10, interferon-inducible protein 10; MIG, monokine induced by interferon-γ; RANTES, regulated upon activation normal T-cell expressed and secreted; MCP, monocyte chemoattractant protein; MIP, macrophage inflammatory protein; MDC, macrophage-derived chemokine; TARC, thymus and activation-regulated chemokine; SDF-1, stromal-cell derived factor 1.

Target cell		Receptor	Chemokine
T cells	T_H1	CXCR3	IP-10, MIG
	T_H2	CCR3 CCR4	Eotaxin, RANTES, MCP-3/4/5 RANTES, MIP-1α, MCP-1, TARC, MDC
	Both	CXCR4	SDF-1α
Eosinophils		CCR1	MCP-3
		CCR2	MCP-4
		CCR3	Eotaxin, RANTES, MCP-3/4/5

vascular endothelium through interactions between cutaneous lymphocyte antigen (CLA) on the lymphocyte with E-selectin on the endothelial cells. The rolling cells are brought into contact with chemokines retained on heparan sulfate proteoglycans on the endothelial cell surface. Chemokine signaling activates the lymphocyte integrin LFA-1, leading to firm adherence to ICAM-1 on the endothelium followed by extravasation, or departure from the blood vessel, into the skin (see Fig. 13.1). Eosinophils migrate into tissues in a similar way, via an interaction between the integrin VLA-4 on the eosinophil and VCAM-1 on the vascular endothelium. Once the lymphocytes and other cells have crossed the endothelium into the dermis, their migration to the focus of inflammation is directed by a gradient of chemokine molecules bound to the extracellular matrix.

If the T cells are activated by specific antigen in the tissue that they enter, they produce chemokines such as RANTES, as well as cytokines such as TNF-α, which activates endothelial cells to express E-selectin, VCAM-1, and ICAM-1. The chemokines act on other T cells to upregulate their adhesion molecules, thus increasing the recruitment of T cells into the affected tissue. At the same time, monocytes and polymorphonuclear leukocytes are recruited to these sites by adhesion to E-selectin. The TNF-α and interferon (IFN)-γ released by the activated T cells also act synergistically to change the shape of endothelial cells, resulting in increased blood flow, increased vascular permeability, and increased immigration of leukocytes, fluid, and protein into the site of inflammation. Thus, a few allergen-specific T cells encountering antigen in a tissue can initiate and amplify a potent local inflammatory response that recruits both antigen-specific and accessory cells to the site.

The case of Tom Joad: the itch that rashes.

Tom was admitted to hospital when he was 2 years old because of his worsening eczema. In the week before admission he had developed many open skin lesions (erosions), increased itching (pruritis), redness (erythema), and swelling (edema) of the skin. The lesions oozed a clear fluid, which formed crusts around them.

Tom had suffered from eczema since he was 2 months old, when he developed a scaly red rash over his cheeks and over his knees and elbows (Fig. 30.3). He was breast-fed until 3 months old, when he was given a cow's milk-based formula. After 24 hours on the formula he started to vomit and to scratch his skin. A casein hydrolysate formula was substituted for milk and he tolerated this well, but as new foods were added to his diet, the eczema worsened. At 9 months old he developed a wheeze and was treated with bronchodilators. At 2 years old he had hives after eating peanut butter.

Tom's mother suffers from hayfever and his father has atopic dermatitis. The family lives in an old house with 20-year-old carpeting. Tom slept on a 10-year-old mattress surrounded by lots of stuffed animals. There were no pets and his parents did not smoke.

During physical examination Tom was evidently uncomfortable and scratched continuously at his skin. His temperature was 37.9°C, pulse 96, respiratory rate 24, blood pressure 98/58 mmHg, weight 12 kg (10th percentile), and height 90 cm (25th percentile). His skin was very red, with large scales, and with scratched and

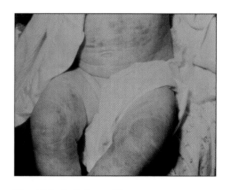

Fig. 30.3 An infant with severe eczema due to atopic dermatitis. Note the reddened and broken skin especially in places, such as over the knees, where it is subjected to continual stretching and stress. Photograph courtesy of S Gellis.

2-year-old boy with severe eczema and family history of allergy

infected lesions on his face, trunk, and extremities. Pustules were present on his arms and legs and thick scales in his scalp. Thickened plaques of skin (lichenification) with a deep criss-cross pattern were seen around the creases on the insides of his elbows and knees. and on the backs of his hands and feet.

A skin culture was positive for *Staphylococcus aureus* and *Streptococcus pyogenes* Group A. Tom was treated with intravenous oxacillin (an antibiotic), antihistamines, topical steroids, and skin emollients such as coal tar. The infection resolved and his skin healed. Laboratory studies during hospitalization showed a white blood cell count of 9600 μl^{-1}, with 41% polymorphonuclear leukocytes, 26% lymphocytes, and 25% eosinophils (normal 0–5%), 13.1 g dl^{-1} hemoglobin, and a hematocrit of 37.2%. The absolute eosinophil count was elevated at 2400 μl^{-1} (normal 0–500 μl^{-1}). Serum IgE was also much elevated at 32,400 IU ml^{-1} (normal 0–200 IU ml^{-1}).

After discharge from hospital, Tom attended the allergy clinic, where he was tested for sensitivity to a range of allergens. He showed a positive type I allergic skin response (see Fig. 16.7) to numerous inhaled allergens including dust mites, mold spores, animal dander, and a variety of pollen. Tom also tested positive in a type I skin response to milk, cod, wheat, egg white, peanut, and tree nuts (cashew, almond, pecan, walnut, Brazil nut, and hazel nut) but had no reaction to rice or soybean. To determine whether any of these foods could be causing his atopic dermatitis, double-blind placebo-controlled food challenges were carried out. He developed hives and wheezing after eating 1 gram of egg white, and hives and eczema after drinking 2 grams of powdered milk, but had no reaction to wheat and cod. Four hours after the milk challenge his serum tryptase level was raised, indicating IgE-induced mast-cell degranulation.

His parents were advised to cover his mattress and pillows with a plastic covering and to remove the carpet and stuffed animals from his bedroom to decrease exposure to dust and mite allergens. Tom had a history of allergic reaction to peanuts, so continued avoidance of peanuts and tree nuts was recommended; he was therefore placed on a diet that excluded milk, eggs, peanuts, and tree nuts.

Tom's eczema improved significantly in response to these environmental and dietary control measures, together with the use of emollients and low-potency topical steroids on his skin. The family reduced the risk of skin irritation by avoiding all perfumed soaps and lotions, double rinsing Tom's laundry, and dressing him in cotton clothes. He did well on this regimen until he was 12 years old when he awoke one day with itchy vesicles on his lower left leg and ankle. The lesions progressed to become painful punched-out erosions, and were diagnosed as herpes simplex infection. He was given the anti-viral drug acyclovir and the lesions resolved. Tom is now 15 years old.

atopic dermatitis confirmed; test for common allergen sensitivity

reduce exposure to allergens

Atopic dermatitis.

Atopic dermatitis (AD) is a chronic itchy inflammatory skin disorder that affects about 10% of children to some extent. It can have a significant impact on the quality of life for the patient and his or her family. Many children with AD develop other symptoms of atopy such as food allergies, asthma, and allergic rhinitis (hayfever).

Atopic dermatitis is characterized by a number of immunologic abnormalities including increased IgE levels and impaired cell-mediated immunity. The importance of IgE-mediated hypersensitivity in the condition is suggested by the following clinical findings: most AD patients have significantly elevated IgE levels; allergens exacerbate AD; and there is usually a personal or family history of atopic disease. Defective cell-mediated immunity is present in up to 80% of patients with AD, as manifested by increased susceptibility to severe viral skin infections, increased susceptibility to chronic fungal infections, and decreased responsiveness to delayed-type hypersensitivity skin testing.

The skin lesions in AD contain a mononuclear cell infiltrate that is predominantly in the dermis (Fig. 30.4), and is composed of activated memory CD4 T cells and macrophages. T cells infiltrating the skin lesions express high levels of cutaneous lymphocyte antigen (CLA), which functions as a skin homing receptor for T lymphocytes by binding to E-selectin. Increased numbers of T_H2 CD4 cells are found in AD, and might be responsible for the raised IgE levels and the weak cell-mediated immunity. Cells infiltrating acute AD skin lesions contain increased levels of mRNA for IL-4, IL-5, and IL-13, which are characteristic of a T_H2 response. Consistent with this, studies have found that allergen-specific T-cell clones from affected skin produce IL-4 but little or no IFN-γ (the latter is characteristic of a T_H1 response). As well as IL-4, IL-5, and IL-3, activated T_H2 cells found in acute skin lesions secrete the cytokines IL-6, IL-10, IL-13, and GM-CSF, and chemokines such as eotaxin and RANTES. These cytokines promote B-cell responses, upregulation of IgE receptors on Langerhans' cells, and downregulation of T_H1 activity and IL-1 receptors on monocytes. The latter effects might account for the lack of delayed-type skin reactivity seen in most patients with AD.

Two of these cytokines in particular have critical roles in the condition. IL-4 increases IgE synthesis by promoting B-cell isotype switching to IgE, and also stimulates the preferential differentiation of T_H2 cells from naive CD4 T cells on encounter with antigen. IL-5 promotes the differentiation and survival of eosinophils, which secrete a range of inflammatory mediators. IL-4 can also activate keratinocytes and fibroblasts, which then secrete the chemokines IL-8 and MCP-1, and the cytokine IL-1, which attracts and activates T cells, and also secrete the chemokines MCP-3 and MCP-4, which attract eosinophils.

Langerhans' cells and macrophages that infiltrate the AD skin lesions have IgE bound to their surface through CD23, a low-affinity receptor for IgE. Langerhans' cells and monocytes can act as antigen-presenting cells, which can process and present the allergen to naive T cells and activate them. The presence of bound IgE on the surface of these cells allows them to concentrate the antigen and renders them more efficient at antigen presentation. IgE is also bound to mast cells in the tissues through the high-affinity IgE receptor FcεRI. Signaling through this receptor after the crosslinking of IgE by antigen leads to the production and secretion of IL-4 and IL-5 by the mast cells, thus further biasing the T-cell response to a T_H2 phenotype.

Treatment of AD is aimed at softening the underlying dry skin and reducing inflammation. AD skin has increased water loss through the epidermis, which reduces its barrier function and facilitates sensitization through the skin. Emollients are the first line of topical therapy for AD because they improve the skin's barrier function. Avoidance of irritants such as soaps and synthetic fabrics, which disrupt this barrier, is crucial in controlling AD. Acute flare-ups of AD require treatment with topical corticosteroid ointments or creams. Anti-histamines can be helpful, especially at night, to control the itching.

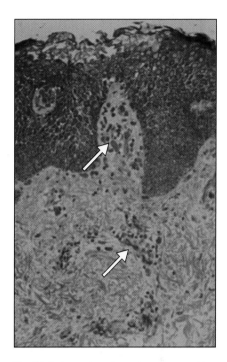

Fig. 30.4 A section through an acute skin lesion from a patient with atopic dermatitis. The section is stained with hematoxylin and eosin. The dermis has been infiltrated by mononuclear cells (arrowed), which are mostly T cells.

Discussion and questions.

__1__ *Topical steroids were effective in reducing the eczema associated with AD. Why?*

Corticosteroids bind to steroid receptors in inflammatory cells such as T cells and eosinophils. The steroid:receptor complex is translocated into the nucleus, where it can control gene expression, including the expression of cytokine genes, by binding to control elements in the DNA. In addition, corticosteroids increase the synthesis of the inhibitor of the transcription factor NFκB, which controls the expression of multiple cytokine genes. One effect is to inhibit the synthesis of cytokines and the release of preformed mediators and arachidonic acid metabolites. Although topical steroids are very effective, excessive or prolonged use of powerful steroids can lead to local skin atrophy.

__2__ *What other immunomodulatory agents might be effective in AD?*

The immunosuppressant cyclosporin A acts primarily on T cells and interferes with the transcription of cytokine genes. The drug binds to an intracellular protein, cyclophilin, and this complex in turn inhibits calcineurin, which normally dephosphorylates NF-AT, a major cytokine gene transcription factor. FK-506, or tacrolimus, is another immunosuppressant with a spectrum of activity similar to that of cyclosporin. Tacrolimus binds to the cytoplasmic protein FK-506-binding protein, and this complex also inhibits calcineurin. Tacrolimus has a smaller molecular size and higher potency than cyclosporin A and, perhaps because of these features, it seems to be effective as a topical formulation.

__3__ *Why did Tom develop an extensive herpesvirus infection?*

Patients with AD have defective cell-mediated immunity, which is required for the control of herpesvirus infections. Cell-mediated immunity involves T_H1 CD4 cells and CD8 cytotoxic cells; patients with AD have selective activation of T_H2 rather than T_H1 cells, as shown by their reduced delayed-type hypersensitivity skin reactions. They also have decreased numbers and function of CD8 cytotoxic T cells. Furthermore, monocytes from patients with AD secrete increased amounts of IL-10 and prostaglandin E_2 (PGE_2). Both IL-10 and PGE_2 inhibit the production of the T_H1 cytokine IFN-γ, and IL-10 also inhibits T-cell-mediated reactions.

__4__ *Atopic dermatitis is described as the 'itch that rashes.' If patients are prevented from scratching, no rash occurs. What is the relationship of scratching to the rash?*

Scratching causes tissue damage that stimulates the keratinocytes to secrete cytokines and chemokines (IL-1, IL-6, IL-8, GM-CSF, and TNF-α). IL-1 and TNF-α induce the expression of adhesion molecules such as E-selectin, ICAM-1, and VCAM-1 on endothelial cells, which attract lymphocytes, macrophages, and eosinophils into the skin. These infiltrating cells secrete cytokines and inflammatory mediators that perpetuate keratinocyte activation and cutaneous inflammation.

5 *Skin infections with staphylococci and other bacteria exacerbate AD. Can you suggest a possible explanation for this?*

The skin of more than 90% of patients with AD is colonized by *Staphylococcus aureus*. Recent studies suggest that *S. aureus* can exacerbate or maintain skin inflammation in AD by secreting a group of toxins known as superantigens, which cause polyclonal stimulation of T cells and macrophages (see Case 7). T cells from patients with AD preferentially express T-cell receptor β chains $V_\beta 3$, 8, and 12, which can be stimulated by staphylococcal superantigens resulting in T-cell proliferation and increased IL-5 production. Staphylococcal superantigens can also induce expression of the skin homing receptor (CLA) in T cells, which is mediated by IL-12. Additionally, nearly half of AD patients produce IgE directed against staphylococcal superantigens, particularly SEA, SEB, and toxic shock syndrome toxin-1 (TSST-1). Basophils from AD patients who produce anti-toxin IgE release histamine on exposure to the relevant toxin. These findings suggest that local production of staphylococcal exotoxins at the skin surface could cause IgE-mediated histamine release and thereby trigger the itch–scratch cycle that exacerbates the eczema.

6 *How do environmental allergens such as dust mites, molds, and animal danders trigger atopic dermatitis?*

Inhaled allergens can enter capillaries in the lungs and thence be carried to the skin in the bloodstream. Allergen binding to Langerhans' cells and mast cells bearing surface IgE in the skin results in cytokine and mediator release, which causes edema and inflammation. In patients with AD and asthma, inhalation challenge with dust mite allergen results in both bronchial and skin reactions in a subset of patients. Skin reactions occurred 1.5–17 hours after the mite allergen inhalation challenge. Sensitization directly through the skin can also occur. A mouse model of atopic dermatitis suggests that sensitization directly through the skin can even result in allergen-induced asthma. In this model, patch application of allergen to the shaved skin of a normal mouse results in an eczematous dermatitis and subsequent allergen-specific airway hypersensitivity such that a single first-time encounter with the allergen by inhalation causes airway hyper-responsiveness typical of the asthmatic state.

INDEX